Behavioral Problems in Geography

Originally published in 1969, Behavioral Problems in Geography unpacks and identifies elements of behavioral models and theories. The book seeks to examine their specific effects on spatial activity and to operationalize some of the concepts previously used in a subjective and descriptive manner. All papers are united by a common concern for the building of geographic theory regarding human behavior. Contributions vary a great deal in their emphasis ranging from philosophy and review, to theorizing and operationalization. Each paper recognizes the importance of examining the behavioural basis of spatial activity. This book will appeal to scholars of geography and psychology alike.

Behavioral Problems in Geography

A Symposium

Edited by
Kevin R. Cox and Reginald G. Golledge

Routledge
Taylor & Francis Group

First published in 1969
by Northwestern University

This edition first published in 2018 by Routledge
2 Park Square, Milton Park, Abingdon, Oxon, OX14 4RN
and by Routledge
711 Third Avenue, New York, NY 10017

Routledge is an imprint of the Taylor & Francis Group, an informa business

© 1969 Kevin R. Cox and Reginald G. Golledge

Publisher's Note
The publisher has gone to great lengths to ensure the quality of this reprint but
points out that some imperfections in the original copies may be apparent.

Disclaimer
The publisher has made every effort to trace copyright holders and welcomes
correspondence from those they have been unable to contact.

A Library of Congress record exists under LCCN: 72027777

ISBN 13: 978-0-8153-7827-3 (hbk)
ISBN 13: 978-1-351-23271-5 (ebk)
ISBN 13: 978-0-8153-7829-7 (pbk)

The William and Marion Haas Research Fund

NORTHWESTERN UNIVERSITY
Studies in Geography

Number 17

BEHAVIORAL PROBLEMS IN GEOGRAPHY: A SYMPOSIUM

Kevin R. Cox and Reginald G. Golledge, Editors

Department of Geography
Northwestern University
Evanston, Illinois
1969

PREFACE

Eight of the papers in this volume were originally presented in a special session entitled "Behavioral Models in Geography" at the annual meetings of the Association of American Geographers in Washington, D. C., 1968. The two additional papers by Harvey and Olsson were solicited to complement the papers presented at the meeting.

The session was designed to expose first drafts of the papers to critical discussion on the part of interested geographers. Formal discussion at the session was provided by Peter R. Gould, Pennsylvania State University, Gunnar Olsson, University of Michigan, Allan Pred,.University of California at Berkeley, and Edward Soja, Northwestern University. Comments made by the discussants and by others who read or heard the papers have proved extremely useful in revisions of the papers.

In the light of the length and number of papers and the limited time available a considerable amount of credit is also due to Forrest R. Pitts of the University of Hawaii for his skillful chairing of the session.

In addition we must also acknowledge those who have been closely connected with editorial phases in the production of this volume. For critical editorial comment on a number of the papers in this volume we would like to thank Dr. Leslie J. King, Department of Geography, Ohio State University. For financial assistance in preparing the papers for the publisher thanks are due to the College of Social and Behavioral Sciences, Ohio State University.

Kevin R. Cox
Reginald G. Golledge
Ohio State University

CONTENTS

Page

EDITORIAL INTRODUCTION:

BEHAVIORAL MODELS IN GEOGRAPHY

Kevin R. Cox
and
Reginald G. Golledge
Ohio State University

Introduction

The papers brought together in this volume are all representative of a growing interest on the part of geographers in
what has been very loosely termed the behavioral approach to
geographical problems. Eight of the papers, since revised,
were originally presented at an arranged session at the annual
meetings of the Association of American Geographers, Washington,
D. C., 1968. The two additional papers were solicited for that
meeting but were not available at that time.

Contributions to this volume vary a great deal in their
emphasis ranging from philosophy and review, to theorizing and
operationalization. Each paper, however, recognizes the importance of examining the behavioral basis of spatial activity.
This is not new of itself, for many exisiting theories and
models in geography have at least an implicit behavioral element in their structure. What is new is the deliberate
attempt to unpack and identify these elements, to examine their
specific effects on spatial activity, and to operationalize
some of the concepts previously used in a subjective and descriptive manner. All papers, therefore, are united by a common concern for the building of geographic theory on the basis
of postulates regarding human behavior. The aim of this brief
editorial comment is to fit the papers into this general context by pointing out the role of behavioral models in

contemporary geography, the types of behavioral model which are relevant, the sources of behavioral postulates and the implications of such models for the operational procedures of the geographer.

The Role of Behavioral Models

In order to anticipate questions regarding the need for the approach adopted in this volume, an initial comment concerning the role of behavioral models in geography is necessary. At a time when geography is undergoing very rapid change in both theory and technique it seems only reasonable to challenge the relevance and necessity of each emerging research theme. Recent papers by Harvey,[1] and Olsson and Gale,[2] and a monograph by Pred,[3] have attempted to answer the questions of need and relevance. For example as Harvey has recently stated, the postulates of geographic theory are in part indigenous and in part derivative: "Indigenous geographic theory may be regarded ...as an attempt to state the laws of spatial form in the specialized languages of geometry or topology, or in the more general form of spatial statistics."[4] He also points out however, that such laws tell one little or nothing about processes, and for this geographic theory must turn to the other social sciences and the postulates about human behavior which they can provide. In order to understand spatial structure, therefore, we must know something of the antecedent decisions and behaviors which arrange phenomena over space. This is the viewpoint stressed in this volume.

Types of Behavioral Models for Geographic Research

A large number of behavioral mechanisms have spatial correlates. This is particularly evident, for example, in those studies which make use of the concept of information flow, where the sender and receiver of the information can be identified as to location and where the efficacy of the mechanism can be related to relative location.[5] It is also evident

that there is a growing number of studies in geography which
focus upon the mental storeage of spatial information.[6]
Mental map studies and action space studies, for example, have
implicit within them some relationship to the movements and
activities which produce spatial structure, and spatial structure cannot be understood without some knowledge of the perception of spatial reality that is retained in the human mind.
Hence the emphasis of several papers in this volume is upon
social and psychological mechanisms which have explicit spatial
correlates and/or spatial structural implications.   In this
sense the work differs from some of the work done in geography
upon environmental perception where the interest has been less
in behavioral mechanisms having spatial correlates and more
upon the measurement of attitudes toward environmental
stimuli.[7]

Preliminary analyses of behavior in space have indicated
two distinct but complementary levels of research.   The basic
level consists of a search for relevant postulates and models
to describe behavioral processes irrespective of the spatial
structure in which the behaviors are found.   In other words
this search is for the rules of choice, movement, and interaction which are independent of the spatial system in which
they are to operate.   Such rules, rather than descriptive
statistics of actual behavior in a system, should provide the
basis for future theorizing.   A second level relates parameters describing actual behavior in an area to specified spatial structures in the same area.   An excellent example of
this level is seen in Hägerstrand's use of mean information
fields, where the parameters of the field are based upon actual
interaction data for the spatial system in question.[8]   Such
parameters, as Rushton has suggested elsewhere, are most likely
related to the spatial structure of interaction opportunities
in the specific spatial system.[9]   Under such circumstances,

behavior in space is constrained by the existing spatial structure and should not be used to explain that structure.

We regard this distinction as a crucial one which must be
borne carefully in mind when reviewing progress in this emerging research theme.   This does not mean to say, however, that
the first level should be developed to the exclusion of the
second level.   Certainly Hägerstrand's work on the diffusion
of innovations was based upon a careful and apparently realistic evaluation of the processes generating the observed spatial
structures in his study areas.   In this way he provided a very
important foundation for other studies in which the interpersonal contacts over space which ignite the spatial diffusion
process, can be predicted independently of existing spatial
structure.   Furthermore studies at the second level, particularly when carried out in several very different spatial contexts, underline the fact that behavioral processes -- regardless of the level of research at which they are discovered --
must eventually be used in conjunction with spatial postulates
to produce meaningful theory.[10]

Sources of Postulates

If theoretically and operationally convincing models of
spatial behavior are to be developed however, a large program
of work lies ahead for geographers in terms of a careful scrutiny of relevant bodies of theory in associated fields.   It
will be obvious to the reader that, in his current attempts to
examine the behavioral basis of spatial activity, the geographer has already drawn heavily on the literature of economics,
psychology, and sociology.   This cross-fertilization is but
further evidence of an increasing cooperation among social and
behavioral scientists that has given considerable returns in
other fields of study.   While it is expected that this cooperative trend will continue, we must not hope for too much
in the way of immediate and substantial returns, for as Harvey

so chasteningly points out, the required scrutiny of other
sciences for relevant material will be long and arduous.
However a start must be made; the papers in this volume provide evidence that some researchers have already delved deep
enough to find interesting and significant contributions.

At present, the behavioral postulates employed in geographic research can be primarily related to three fields of
inquiry; economics, sociology, and psychology. Of these, the
longest and most enduring influence has been exerted by economics. This influence is reflected in the essay on migration
and place utility by Brown and Longbrake, who employ a linear
programming model with classic system minimization criteria in
order to allocate migrants to locations in a competitive housing market. However, as with other papers, the authors draw
on more than just the one field. The linear programming
model is used to operationalize the concept of place utility
-- a concept originally developed from joint knowledge of psychology, geography, and economics. Similarly, Rushton takes
the ideas of revealed preference from micro-economics and uses
scaling models developed by psychologists to give the concept
a spatial interpretation in terms of visits by dispersed rural
consumers to urban places. Golledge's examples of possible
uses of learning theory in geography also draw on both economic
and psychological concepts of consumer behavior in an attempt
to understand spatial choice and movement.

Sociology has contributed only slightly less than economics
to the behavioral mechanisms employed in geographic research.
The pioneer in self-conscious resort to the postulates of
sociology and sociometry was Hägerstrand who, in the 1950s,
worked upon the relationships between informal social relationships on the one hand and the spatial diffusion of innovations
and migration on the other hand.[11] Examination of social
relationships over space has also allowed the explication of

discrepancies between distirbutions predicted by assumptions
of economic man and real world distributions.[12]

The use of sociological postulates, however, while explaining some of the discrepancies between the real world and a
spatial theory based on economic man assumptions, is bound to
be less than perfect in the sense that information is rarely
received in the same form in which it is transmitted. Intervening between the sending of information and the decision to
locate are *perceptions* of information. This was possibly
realized in the work of geographers on map transformation.[13]
It is only more recently however that emphasis has switched
from assumptions of perceptual distortion of information to
actual inquiry into the nature of perceptual distortion of
space and the way in which it affects locational decision
making. On a more limited scale, recourse has also been made
to psychological research on nonmetric measurement and learning theory for assistance in explaining spatial behavior.
Examples of this trend are noted in the papers by Rushton and
Golledge.

While singling out economics, sociology, and psychology as
having made significant contributions to the papers in this
volume, however, it is to be hoped that geographers will
examine all the behavioral sciences in order to obtain meaningful concepts and postulates for the refashioning of geographic
theory. Clearly such theory will need to be constantly revised
to accomodate revisions made by researchers in other fields.

The Symposium Papers

The papers in this volume represent three approaches which
are symptomatic of the state of the art. The papers by
Olsson and Harvey, for example, are both overviews and philo-
sophical. Their aim is to review some of the major trends in
behavioral research in geography, to point up the problems

associated with different lines of research, and to relate
this work to other research in the field. The other eight
papers can be divided into those that are process oriented
per se, and those that are more cross-sectional in scope and
concentrate on operationalizing key concepts. Process oriented
papers include those by Wolpert and Ginsberg, Golledge, Cox
and Reynolds. The processes involved are decision making,
learning, and information flow in a social relations network
respectively. The concepts emphasized by Brown and Longbrake,
Rushton, Stea and Morrill and Earickson are place utility,
space preference, mental mapping, and least-effort allocation.
There is no need to expand further on the origin of these processes and concepts, as each author indicates clearly the
sources of his inspirations.

Implications and Future Research Directions

It is of course too early to judge the effects of behavioral
work in geography on the field as a whole. However, some
short term effects are already obvious, and speculations can
be made concerning the long term effects. One short term
effect is that more geographers are becoming interested in
theory and the problems of premature operationalization of
theory are being realized.[14] The procedure (classic in the
early 1960's) by which certain hypotheses -- often isolated
from and logically inconsistent with each other -- were defined
and tested, seems to be giving way to a phase in which testing
and identification of parameter values are deferred until adequate theory is formulated. The emergence of some stimulating
conceptual frameworks in the last three or four years is
healthy evidence of this.[15]

A second short-run effect has been to underline the arti-
ficiality of the boundaries between the present systematic
fields of geography. The learning models stressed by Golledge
as applicable to shopping center patronage and agricultural

marketing decisions are also applicable to agricultural land
use decisions[16] and indeed to any spatial diffusion problem.[17]
The mental map literature has implications not only for migration but also for an understanding of urban travel movements,[18]
while the notion of information flow is applicable not only to
the spatial diffusion of innovations[19] and the study of voting
response surfaces[20] but also to economic geography.[21]   Indeed,
it seems more than likely that when geography is rewritten in
the future at least a twofold division will be apparent: a
section on spatial structure and a section on the processes
responsible for the observed structures.   It is in the processes section that behavioral models should appear in a domi-
nating position.

Over a longer time period implications of a technical
character become apparent in terms of both the appropriate
models and the data with which to test the models.   As far as
quantitative models are concerned, two needs seem to be emerging.   There is firstly a need for models which mirror behavioral processes in a way which, for example, the traditionally
used gravity model does not.   Rushton's model of space preference comes very close to this theoretical ideal in that it
is an attempt to mirror the comparison and ranking of interaction opportunities apparently underlying the decisions which
determine spatial behavior.   The advocacy of stochastic models
for studies of spatial behavior may be viewed in a similar
light.   Second, there is a need for operational models which
can areally generalize on the basis of small amounts of individual data.   The usual constraints of individual data gathering imply a focus upon a small-scale spatial system;   the
interest of the geographer however is not only at this scale
but beyond it.   For this reason spatial simulation studies of
the type which Hägerstrand and Morrill in particular have proposed assume a magnified significance.[22]

Given models which can mirror the behavioral processes
symbolized in the emerging theoretical constructs of behavioral
geographers there is clearly a need for individual data with
which those models can be operationalized. Such data should
refer to the same individual at a sequence of spatio-temporal
locations.[23] The question of future data requirements is too
complex to be discussed here. However, two specific comments
can be offered. First, since geographers still deal with
populations much more than they deal with individuals it seems
reasonable to press for more discussion of the problem of data
aggregation. Second, it appears that the low levels of explanation achieved by testing many existing models are at
least partly explained by inadequate selection of sample populations. It is unreasonable to expect an optimizing model to
explain actions which are less than optimal. There seems to
be a need therefore to investigate the possibility of stratifying populations on the basis of their achievements, and to
build more complex models which recognize disparities in
populations.

Concluding Comments

Clearly the data and model considerations discussed in the
latter part of this introductory note cannot be divorced from
those theoretical aims of behavioral work in geography which
we discussed earlier. A recent paper by Coleman[24] has attempted to place the relationships between theory, data, and operational models in perspective and the points which he makes
have important implications for future work in geography.
Coleman draws a contrast between two types of approach in the
building of models of social behavior. On the one hand there
are those models in which the theoretical basis is relatively
meager and in which the analysis must supply the missing information about behavioral parameters; such models require heavy
inputs of data in order to say anything useful about the system

10

to which their data pertain and due to their reliance upon
system-specific data their degree of generality is often weak.
On the other hand, there are models which have a rich theoretical basis and which therefore require little real-world data
to describe behavioral propensities; instead such propensities
are dictated by the theory. Data input for these models is
correspondingly small and the general applicability of the
model is greatly enhanced. As Coleman himself states: "What
is necessary...is a more economical model: one which can
derive the probabilities of action *in particular circumstances*
(our italics) from a more general principle. Only in this
way is it possible to escape from the enormous appetite for
data that a model for a complex system has. Not surprisingly
this aim of economy of data is the justification for theory in
any area. And the introduction of a general principle from
which an individual's action can be derived is exactly the
introduction of theory."[25]

   Coleman's comment is an apt point at which to terminate
these introductory comments underlining as it does the significance of studies of the first level which we defined earlier:
i.e. studies involving the identification of postulates and
models to describe behavioral processes, irrespective of the
spatial structure in which the behaviors are found. Not only
are such studies economical of individual data but only with
such an ultimate focus can geographers hope to produce theory
with the degree of generality to which they aspire.

11

NOTES

1. David W. Harvey, "Behavioral Postulates and the Construction of Theory in Human Geography," *Bristol Seminar Paper, Series A, No. 6* (1967), (to be published in *Geographica Polonica*, 1969).

2. Gunnar Olsson and Stephen Gale, "Spatial Theory and Human Behavior," *Papers and Proceedings of the Regional Science Association*, Vol. XXI (1968), 229-242.

3. Allan Pred, *Behavior and Location. Foundations for a Geographic and Dynamic Location Theory, Part I.* (Lund: Gleerup, 1967).

4. Harvey, *op. cit.*

5. The spatial diffusion literature contains numerous examples of such models: see, for example: (i) Torsten Hägerstrand, "On the Monte Carlo Simulation of Diffusion," *European Journal of Sociology*, VI (1965), 43-67; (ii) Lawrence A. Brown, *Diffusion Dynamics: A Review and Revision of the Quantitative Theory of the Spatial Diffusion of Innovation* (Lund: Gleerup, 1968a).

6. The mental-map literature is an outstanding case of such studies: see, for instance, Peter R. Gould, "On Mental Maps," *Michigan Inter-University Community of Mathematical Geographers*, Discussion Paper No. 9 (1966).

7. Consider, for example: (i) Robert W. Kates, *Hazard and Choice Perception in Flood Plain Management.* University of Chicago: Department of Geography, Research Paper No. 78 (1962); (ii) Thomas F. Saarinen, *Perception of Drought Hazard on the Great Plains.* University of Chicago: Department of Geography, Research Paper No. 106 (1966).

8. Hägerstrand, *op. cit.*

9. Gerard Rushton, "Analysis of Spatial Behavior by Revealed Space Preference," *Annals of the Association of American Geographers*, Vol. 59 (1969, forthcoming).

10.   See, for example: (i) Leslie J. King, "The Analysis of Spatial Form and its Relation to Geographic Theory," *Annals of the Association of American Geographers,* (1969, forthcoming);   (ii) Harvey, *op. cit.*

11.   Hägerstrand, *op. cit.* ;   also Torsten Hägerstrand, "Migration and Area," in David Hannerberg, Torsten Hägerstrand, Bruno Odeving (eds.), *Migration in Sweden* (Lund: Gleerup, 1957).

12.   An interesting example of this can be found in Julian Wolpert, "The Decision Process in a Spatial Context," *Annals of the Association of American Geographers,* Vol. 54 (1964), 537-558.

13.   See Waldo R. Tobler, "Geographic Data and Map Projections," *Geographical Review,* Vol. 53 (1963), 59-78;   also see Hägerstrand (1957), *op. cit.*

14.   There is, however the possibility that, as King has pointed out in discussion of the behavioral trend, the interest in theory prompted a concern for behavioral postulates rather than the other way around.   See King, *op. cit.*

15.   Examples include: (i) Lawrence A. Brown, *Diffusion Processes and Location* (Philadelphia: Regional Science Research Institute, 1968b);   Pred, *op. cit.;*   Rushton, *op. cit.;* Julian Wolpert, "Behavioral Aspects of the Decision to Migrate," *Papers and Proceedings of the Regional Science Association,* Vol. 15 (1965), 159-172.

16.   Peter R. Gould, "Wheat on Kilimanjaro: The Perception of Choice within Game and Learning Theory Frameworks," *General Systems Yearbook,* Vol. 10 (1965), 157-166.

17.   Hägerstrand (1965), *op. cit.* conceptualizes the adoption of innovations as the result of a learning process.

18.   R.G. Golledge, R. Briggs and D. Demko, "Configurations of Distance in Intra-Urban Space," (unpublished manuscript, Department of Geography, Ohio State University, 1969).

19.   Brown (1968a), *op. cit.*

20.   See, for instance: (i) David R. Reynolds and J. Clark Archer, "An Inquiry into the Spatial Basis of Electoral Geography," *Department of Geography, University of Iowa,* Discussion Paper No. 11 (1968);   (ii) Kevin R. Cox, "The Voting Decision in a Spatial Context," in C. Board, R.J. Chorley and P. Haggett (eds.), *Progress in Geography* (London: Edward Arnold, 1969).

21. Gunnar Tornquist, "Flows of Information and the Location of Economic Activities," *Geografiska Annaler,* Vol. 10, Ser.B, No. 1 (1968), 99-107.

22. Hägerstrand (1957, 1965), *op. cit.;* Richard L. Morrill, "The Negro Ghetto: Problems and Alternatives," *Geographical Review,* Vol. 55 (1965), 339-362.

23. Olsson has drawn attention to this problem in his contribution to this volume.

24. James S. Coleman, "The Use of Electronic Computers in the Study of Social Organization," *European Journal of Sociology,* Vol. 6, No. 1 (1965), 89-107.

INFERENCE PROBLEMS IN LOCATIONAL ANALYSIS

Gunnar Olsson

The University of Michigan

A. Introduction

Contributions to this and other recent volumes indicate
that some quantitative geographers have shifted their attention from the modelling of large-scale aggregates to studies
of group and individual behavior.   As a result, the earlier
stress on the geometric outcome of the spatial game has less-
ened in favor of analyses of the rules which govern the moves
of the actors who populate the gaming table.   Thus, the new
studies aim at a better understanding of those cause and effect
relationships which are relevant to the decision makers themselves, i.e. to those whose actions eventually will determine
the success of various planning programs.   With such pragmatic planning ideas in mind, the behaviorists wish to complement the traditional work in quantitative geography by establishing explicit linkages between individual behavior and
spatial patterns.   Restated and simplified, the behavioral
approach suggests a different solution to the geographical
inference problem of form and process;   while the spatial
analyst attempts to infer individual behavior from knowledge
of a given spatial pattern, the behaviorist argues for reasoning the other way around.

To assess the epistemological merits of the suggested
solutions to the geographic inference problem, the initial
purpose of this paper is to discuss current location theories
from the viewpoint of the philosopher of science. The secondary purpose is to investigate how well the ideal approach of
the behaviorists outlined in the first half of the paper
actually compares with their operationalized models. The
latter investigation is prompted by the realization that it is
one thing to make attractive methodological statements but
quite another to translate such predilections into testable
formulations.

B. The Truth Status of Geographic Theory

A major proposition of the Vienna Circle is that scientific
statements become lawlike by being logically consistent and
empirically true, i.e. by being acceptable in terms of both
syntax and semantics. A theory may then be defined as a set
of deductively connected laws.[1] It follows that once the
syntax of a specific theory is accepted as true, its additional
value can be assessed in terms of semantics or by its ability
to predict empirical events. To test a statement derived
from a theory is therefore to seek the instantiation of what
is presently considered a law.[2] An important vehicle in this
verification procedure is the notion of correspondence rules
by which the calculus of the theory can be interpreted in
terms of real world observations, or, if an alternative view
is adopted,[3] the calculus of the real world can be interpreted
in terms of the theory. In case the correspondence rules
cannot be specified or if their application indicates differences between the theoretical and observational languages,
then the truth status of the theory is in question. Conversely, a theory is held to be true when all its extra-logical
terms have factual reference.[4]

Even though geography rarely has been discussed within the rather stringent framework of epistemology,[5] it may still be suggested that predictions based on geographic theory tend to be dubious. This is certainly the opinion of Pred, who argues that the breakdown of location theory is due to discrepancies between the motives which govern the decisions of the actors in the theory and the actors in reality.[6] Others have conveyed essentially the same message about the validity of existing theories but in a lower key and relating it to other causes. For instance, Curry has criticised Lösch's notion of the unbounded plain so severely that he advocates that "little remains of existing theory to allow its refashioning."[7] Dacey, finally, is less definitive, despite his claim that it is "inconceivable that any pattern of central places corresponds exactly to the specified geometry."[8]

There can be many reasons why theoretical predictions do not sufficiently agree with empirical observations. One reason relates to the aspiration level or the degree of generality at which a particular theory aims. This observation draws on the fact that theories derived within a hypothetico-deductive system by definition can be ordered into a hierarchical structure of statements.[9] An important characteristic of those structures is that axiomatic statements on one level may be testable theorems of a higher level theory. This suggests that a particular theory should not be rejected simply because it contains unrealistic axioms and therefore may provide unsatisfactory high level predictions. At the same time, it should be noted that the possibility of syntactical mistakes cannot be ruled out until the theory has been completely axiomatized. It reflects some of the biases of the work in theoretical geography that the only attempts at axiomatization relate to the geometric properties of Christaller's and Lösch's theories.[10]

The previous arguments suggest that the low predictive
power of geographic theories perhaps may be due to weak linkages in the interlocking system of hierarchically ordered
statements.   If this is true, then it is difficult, perhaps
impossible, to say anything conclusive about the validity of
existing constructs.   Since the theories have not yet been
fully formalized, it is uncertain exactly which aspiration
level is being sought.   The root of the uncertainties is supposedly in the employed axioms or perhaps rather in ambiguities
introduced when these were provided with semantical meaning
and turned into assumptions.   It is the purpose of the next
section of the paper to discuss the validity of this supposi-
tion.

C. The Assumptions of Geographic Theory

It is sometimes helpful to make a superficial distinction
between the spatial and the behavioral axioms of location
theory.   More exactly, the former postulates relate to the
properties of the area over which the actions occur, while the
latter concern the motives and behavior attributed to the actors themselves.   The behavioral assumptions are usually the
same as those of the theory of the firm, while the spatial
assumptions commonly are those of the unbounded homogeneous
surface.   By combining spatial and behavioral postulates,
theorems like the hexagonal arrangement of central places may
be derived.

To point out that the assumptions of location theory are
unrealistic is almost trite.   It is far more important that
the lack of realism becomes critical only above a certain
aspiration level, i.e. when a higher level hypothesis is re-
futed by observations which would not refute a lower level
hypothesis.[11]   Taking the pragmatic view of many economists,[12]
it is enough, therefore, to decide whether the assumptions

18

lead to sufficiently accurate predictions for the purpose at
hand -- it would be foolish to stop making everyday predictions
on the basis of Newton's laws only because of Einstein's subsequent work. Likewise, provided the location analyst is content with devoting himself to pattern analysis and interpretive
descriptions of large-scale spatial regularities, he may possibly be satisfied with existing constructs. If, on the other
hand, the aspiration level were changed to include analyses
also of those micro-units, groups or individuals whose actions
give rise to the large-scale regularities, then the situation
would be different. The reason is, of course, that the
axioms in the first case have become testable theorems in the
second. There is little doubt that on this new aspiration
level, the axioms of the traditional theory are unacceptable.[13]
Instead, the essential problem seems to become that of understanding the internal structure of goal conflicts and adaptive
systems. It is for reductions to this level of explanation,
i.e. to the level at which inner and outer environments are
treated as adapting to one another,[14] that some behaviorally
oriented geographers are striving.

The question now arises as to whether the imagined new
breed of geographer really is new or whether it is only the
old one dressed up as Hans Christian Andersen's Emperor. It
would probably seem so to geometrically inclined students like
Hudson, who recently claimed that "one central problem of geographic theory is that of relating individual behavior to that
of (spatial) distribution."[15] Few would object. The difference between the spatial and behavioral approaches to this
geographic inference problem becomes clear, however, when
Hudson later refers to a "given system of nodes (within which)
the individual must acquire a set of spatial relations so as
to navigate ... in an efficient manner." Thus, the prime
concern is not to derive a spatial pattern from axiomatized

behavior but rather to make inferences about behavior from the
knowledge of spatial patterns.

Since there are no clearcut one directional cause and
effect relationships between geographic form and process, the
spatial analysis approach of Hudson and others will almost
certainly provide valuable insights. Nevertheless, it is
significant that most geographic geometricians have found it
necessary to tamper with the classical spatial assumptions and
work in transformed non-Euclidean space.[16] Most importantly,
the choice between different assumptions is conceived more as
a matter of expediency than as a problem involving explicit
statements of aspiration level or purpose. This view has been
stated most succinctly by Tobler, who asserts that the theory
can be made "more realistic by relaxing the assumptions, but
(that) this generally entails an increase in complexity."
Difficulties arise, however, when this proposition later is
being executed through the removal of "the differences in geographic distribution by a modification of the geometry or of
the geographical background."[17]

The terms "geographic distribution" and "geographical
background" may convey the impression of areal variations in
physical landscape and population densities, i.e. of phenomena
which have nothing or very little to to with individual or
group behavior. Judging from the rest of the paper, however,
Tobler must have something more far reaching in mind. Thus,
the idea is illustrated by references both to the logarithmic
migration maps of Hägerstrand[18] and to cartograms like 'A New
Yorker's Idea of the United States of America', i.e. to maps
explicitly derived from the theory of cognitive behavior.
This means, of course, that the aspirations in fact have been
increased to a level where the spatial axioms take on the role
of testable theorems derivable from the theory of cognitive
behavior.

What is crucial in the comments on map transformations is not that people seem to do something they claim not to be interested in.   Instead, it is the consequences that the approach has for the testing of classical location theory.   More exactly, it remains to be seen how the data underlying the estimation of transformation functions can be separated from the empirical information with which the theoretical predictions are to be compared;   clear specifications of the employed correspondence rules should help to clarify this issue.   Even then, however, it may be questioned whether the approach actually is as expedient as sometimes suggested.   Speaking exactly to this point, Curry has recently observed that "if we could gain the level of sophistication necessary to transform, we could probably write theory in terms which would not require it."[19]

In short, it appears unclear which exact linkages in the alleged chain of deductive reasoning suggested that map transformation would be a valid approach.   In less clever accounts than the ones by Tobler, it even seems that large-scale data occasionally have been used as a basis for inferences about small-scale behavior.   It is tempting to speculate that such logical peculiarities stem either from the traditional reliance on the map as a given, or from the almost metaphysical belief that the same model can be applied to both physical and human phenomena.[20]   More importantly, though, the previous discussion suggests the spatial postulates of location theory to be special cases of behavioral theorems;   the questions asked by the spatial and behavioral analysts therefore tend to belong to different scientific aspiration levels.   This suggestion supports the common sense conclusion that the reliability, explanatory power, and the potential planning applications of any social science theory depend on its treatment of individual and group behavior.

Since the behavioral assumptions of location theory recently have been treated elsewhere, it now seems superfluous to
discuss those in the same detail as the spatial assumptions.[21]
Likewise, it is rather pointless to elaborate on the lack of
realism in terms of actual decision processes. Notwithstand-
ing, it is possible that the postulates still will lead to
sufficiently good results, provided that only large-scale predictions are aspired. If, on the other hand, the goal is to
understand the finer workings behind large-scale regularities,
then it is doubtful whether the traditional approach with its
firm grounding in classical utility theory and normative economics will provide reasonable explanations. Examples of
other, supposedly more fruitful approaches include Hägerstrand's
ideas about information, diffusion, and migrations, Curry's
notions of shopping lists, inventories, queueing, and central
places, Wolpert's work on the spatial attributes of stress and
goal conflicts, and Pred's explorations into the behavioral
matrix.[22] The obvious conclusion is that large-scale patterns should be deduced from explicit statements about individual behavior rather than the other way around. More specifically, spatial patterns should be viewed as reflections of
habits and institutionalizations which in turn can be accounted
for by individual decisions governed by continuous learning
processes.

The attempts to derive spatial patterns from realistic
assumptions about individual or group behavior are appealing
because they aim at higher level explanations. Occasionally,
however, some issues relating to the axioms, theorems, and
aspiration levels still remain unclear. For instance, Curry
has recently suggested that the problem of writing theory is
to obtain postulates which are not so directly linked to the
final results that added insight is not gained.[23] Provided
this means that one should not use the same data sets for

estimating parameter values as for testing theoretical predictions, or that logical deductions may lead to the discovery of new types of scientific facts, then there can be no disagreement. If, on the other hand, the term "gaining added insight" refers to something less tangible, then the statement becomes less clear. This is particularly so in view of the tautological nature of the deductive method.[24]

D. Inference from Geographic Models

The discussion thus far has centered on some epistemological issues relevant to the geographic inference problem of how to connect spatial patterns and human behavior. It has been proposed that this basic problem of form and process can be approached from two basically different directions. Thus, one may arrive at conclusions about individual behavior through analyses of given spatial patterns, or one may draw conclusions about spatial patterns from detailed knowledge of individual behavior. Both approaches involve difficult inference problems, some of which are related to the choice of axiomatic system or aspiration level, while others are embodied in the lack of one-directional cause and effect relationships. Given this situation, continued epistemological reevaluations of locational analyses seem mandatory.

Evidence that such reevaluations are important is provided by the fact that even the very superficial comments of this paper have helped to isolate some attractive features of the behavioral approach. The question therefore arises as to whether the logically appealing syntax can be matched by meaningful semantics. To illuminate this question, the discussion will now turn to a review of some models recently employed by students frequently associated with the behavioral school. It should be noted that only operationalized and tested formulations will be treated; to extend the comments into suggested, hypothetical and nontested constructs would not add anything substantive to the comments already made.

As an introduction to the model review, it may be helpful to rephrase the geographic inference problem in terms of large-scale patterns with small variances and small-scale processes with large variances. It follows that large-scale systems, i.e. systems from which portions of the internal variation has been filtered out, tend to be more deterministic, while small-scale systems are more probabilistic. For this reason, the efforts of rewriting well tested deterministic models like the gravity, regression, and rank size formulations in probability languages becomes interesting from the inference point of view. These attempts to get around rather than solve the problem have been discussed in detail elsewhere,[25] and it is therefore now sufficient to add a reference to some subsequent works on the entropy concept by Wilson.[26]

Provided the same correspondence rules apply in both cases, the rewriting of classical models in probability terms will make large-scale regularities interpretable and derivable from explicit and quasi-realistic assumptions about individual or group behavior. However, the value of these reformulations should not be overestimated; the fact that the final results have been arrived at via another route has not appreciably changed the character of the models. As a consequence, the regulating forces remain deviation counteracting rather than deviation amplifying.[27] On the practical level, however, the translation of the same model to another language with important syntactical and semantical differences may have considerable utility. Thus, it is a common problem in planning situations to determine the scale below which probability techniques must be used; the current discussion of the efficacy of population planning as compared to family planning offers an excellent case in point.[28]

A related but methodologically very different answer to
the problem of connecting large-scale spatial regularities
with small-scale generating mechanisms is provided by the Monte
Carlo simulation technique.   Thus, the main characteristic
of the simulation technique is that evolving spatial patterns
are viewed as resulting from an interplay between deterministic
and random factors.   More specifically, the general development is determined by distance functions translated via the
relative frequency interpretation of probability into the
operational form of mean information fields, while the exact
development is influenced by a large number of chance factors,
operationally represented by the drawing of random numbers.
This means, of course, that the resulting patterns may be
viewed as a set of 'regulated accidents', in which chance and
contingency have played their game with laws of nature and
human behavior.

It has sometimes been suggested that conceiving reality
as the result of regulated accidents is paradoxical because
the observable events are said to obey the laws of chance,
while the underlying probabilities in themselves obey some
causal law.[29]   On the other hand, the same observation has
been extended into the epistemological tenet of complementarity.   The issue can be discussed either in terms of axioms and
theorems as in the first part of this paper, or in terms of
the idea that causal laws at one level of aggregation normally
result from averages of statistical behavior at a deeper level,
which in turn can be explained by deeper causal behavior, and
so on indefinitely.[30]

The diffusion work after Hägerstrand[31] has focused almost
entirely on refining details of the original model.   With
few exceptions, the efforts have been devoted to the construction of specialized computer programs,[32] experiments
with different mathematical distance functions,[33] and the

derivation of biased or unbiased mean information fields,[34]
while little has been done with the more basic issues of
testing and interpretation of underlying theory and functional
relationships. This means that most diffusion students --
Hägerstrand himself not included -- in fact have neglected the
behavioral approach to the geographic inference problem.
Thus the employed procedures involve implicit reasoning from
the large-scale regularities of the mean information field to
the behavior of the individuals as this is governed by the
random number matrix. It follows that more insight may be
gained through detailed experiments with different parameter
values based on observed systematic spatial and temporal variations in resistance, distance sensitivity, communication networks, etc. Proceeding to the testing of subsequent model
generations, this suggests that more attention should be paid
to sensitivity analyses and less to evaluations of spatial
end products.

In practice, spatial applications of the Monte Carlo
technique rely almost exclusively on large-scale aggregate
data, which then are treated as the joint product of deterministic and random variables. As a consequence, inferences
about individual behavior can be made only indirectly via a
reasoning from spatial patterns to generating mechanisms.
The same characterization generally applies to the cell counting technique,[35] even though Dacey in some of his county seat
models has attempted to deduce spatial distributions from
explicit assumptions about underlying processes.[36]

The inferential problem involved in the cell counting
technique is best illustrated in studies employing the negative binomial distribution. It is well known that this distribution can be generated in at least six different ways,
some of which are complete opposites. For instance, a spatial point pattern can be described by the negative binomial

if it consists of randomly located clusters generated through
a two-stage diffusion process in which the parent points have
been distributed randomly over the area, and the secondary
points have been assigned among the initial nuclei independently
of one another but in such a fashion that the growth over
time is logarithmic. The opposite to this generating mechanism is the urn scheme for heterogeneous Poisson sampling,
according to which a negative binomial may be obtained for
the total area, provided the area can be divided into regions
within which the points have been randomly distributed but in
such a manner that the mean number of points per cell varies
between the regions according to the gamma function. The
obvious conclusion is that very little can be said about
generating mechanisms solely on the basis that the morphology
of a point pattern can be described by the negative binomial.[37]

It could perhaps be tempting to conclude from the discussion of the negative binomial that it is possible to reason
from process laws to morphological laws but not in the other
direction. However, not only are the causal linkages in
geography too intricate to allow such a conclusion, but it
may also be shown mathematically that at least one single
mechanism -- that of space filling -- can give rise to either
clustered, random or regular spatial patterns. In such
situations of conflicting results, one solution is to fit the
same data to another set of distributions connectable with
only one of the previous interpretations.[38] Despite its
value in the special case, this approach is clearly more expedient than elegant. It will hardly bring the solution of
the geographic inference problem much closer.

To varying degrees, the models discussed have started
from a given spatial pattern and then proceeded to indirect
inferences about generating processes and individual behavior.
This conclusion has been possible to substantiate only because

the cited models have been refined to the extent of operationalization and empirical application. Unfortunately, the rest of the model work in behavioral geography still awaits rigorous testing and does therefore not permit the same degree of conclusiveness. On present evidence, however, it is not unlikely that current theoretical explorations will end up with testable formulations which are based on the traditional approach to the geographic inference problem rather than on a new logically more attractive one.

The suspicion that techniques for generating spatial patterns from individual behavior still may be far away, relates closely to the attempted adaptations of classical learning theory to the geographers' need.[39] Thus, the focus has been more on how individuals learn to act efficiently in an existing spatial system than on how their actions cause existing spatial patterns to change. Basically, the same holds for the notion of subjective preference functions[40] and for most studies of mental maps.[41] Ignoring the extremely thorny measurement problems as being beside the point in the present context, the most interesting property of the latter studies is their amenability to trend analysis. Perhaps it is on this level that the relationships between the allegedly new approaches and the traditional work in spatial analysis become most evident; trend surface analyses and map transformations appear in fact to be the dual of one another.

E. Summary and Conclusions

This paper was based on the premise that the limited predictive power of geographic theories is due to a preoccupation with spatial patterns and a neglect of small-scale generating processes. The paper has attempted to evaluate this premise by comparing the spatial and the behavioral approaches to the geographic inference problem of form and process. To

establish some general guidelines, attention was first given
to the overlaps between geography and the philosophy of the
social sciences. The subsequent conclusion was that the
behavioral axioms of location theory belong to a higher level
of the hierarchical structure of the hypothetico-deductive
system than do the spatial axioms. As a consequence, the
behavioral approach can provide more detailed explanations and
is therefore preferable, particularly if the findings are to
be extended into planning applications.

It is one thing, however, to isolate attractive methodological approaches and quite another to translate these predilections into operational and testable models. In order to
assess what has actually been achieved rather than merely
talked about, the latter half of the paper reviewed a number
of models recently used by students more or less identified
with the behavioral school. More specifically, attention was
given to the rewriting of deterministic models in probability
terms, the use of Monte Carlo simulations, the cell counting
technique, and the geographical amendments to psychological
learning theory. It was found that practically all studies
had started from given spatial patterns and then proceeded to
indirect inferences about generating processes and underlying
human behavior. Although suggestions about alternative and
epistemologically more attractive approaches do exist, it has
been difficult to find cases where such models actually have
been applied to empirical data. Recalling the positivists'
quest for combinations of logical consistency and empirical
truth, this leaves the assessor bewildered. On the one
hand, it is possible to point to a number of low order spatial
formulations with considerable empirical reliability. On the
other hand, one may imagine some logically attractive behavioristic formulations which unfortunately still await empirical
evaluations.

The final conclusion must be that the behavioral approach to quantitative geography may or may not alter the current rather peculiar state of the art.   Speaking *for* improvement is the growing recognition of studies from quantitative psychology and non-normative economics as well as the epistemological bases of most behavioral work.   Speaking *against* substantial and quick change is the existence of multi-directional causal relationships as well as the shortage of suitable highly disaggregated data.

30

NOTES

1.   Merle B. Turner, *Philosophy and the Science of Behavior* (New York: Appleton-Century-Crofts, 1967), 226; May Brodbeck, *Readings in the Philosophy of the Social Sciences* (New York: Macmillan, 1968), 583.

2.   Thomas S. Kuhn, *The Structure of Scientific Revolutions* (Chicago: University of Chicago Press, 1962).

3.   Henry Margenau, *The Nature of Physical Reality* (New York: McGraw-Hill, 1950); Norwood Russell Hanson, *Patterns of Discovery* (Cambridge: Cambridge University Press, 1958).

4.   Carl G. Hempel, *Aspects of Scientific Explanation* (New York: Free Press, 1965), 217-222.

5.   For notable exceptions see: Fred Lukermann, "On Explanation, Model, and Description," *Professional Geographer*, Vol. 12 (1960), 1-2; Fred Lukermann, "The Role of Theory in Geographical Inquiry," *Professional Geographer*, Vol. 13 (1961), 1-5; Reginald Golledge and Douglas Amedeo, "On Laws in Geography," *Annals of the Association of American Geographers* (1968), 760-774; Gunnar Olsson, *Distance, Human Interaction, and Stochastic Processes: Essays on Geographic Model Building* (Ann Arbor: University of Michigan, 1968); David Harvey, *Explanation in Geography* (London: Edward Arnold, forthcoming).

6.   Allan Pred, *Behavior and Location. Part I* (Lund: Gleerup, 1967).

7.   Leslie Curry, "The Geography of Service Centers within Towns: The Elements of an Operational Approach," in Knut Norborg (ed.), *Proceedings of the I.G.U. Symposium in Urban Geography* (Lund: Gleerup, 1962), 33.

8.   Michael F. Dacey, "Imperfections in the Uniform Plane," *Discussion Papers of the Michigan Inter-University Community of Mathematical Geographers*, No. 4 (1964), 1.

9. Richard Bevan Braithwaite, *Scientific Explanation* (New York: Harper, 1960).

10. Michael F. Dacey, "The Geometry of Central Places," *Geografiska Annaler*, Vol. 47, Ser. B (1965), 111-124.

11. Braithwaite, *op. cit.*

12. Milton Friedman, *Essays in Positive Economics* (Chicago: University of Chicago Press, 1953).

13. Herbert Simon, "Theories of Decision-Making in Economics and Behavioral Science," *American Economic Review*, Vol.49 (1959), 253-283; Richard M. Cyert and James G. March, *A Behavioral Theory of the Firm* (Englewood Cliffs, N.J.: Prentice-Hall, 1963).

14. Herbert Simon, *The Sciences of the Artificial* (Cambridge, Mass.: M.I.T. Press, 1969).

15. John Hudson, "A Model of Spatial Relations," *Geographical Analysis*, Vol. 1 (1969).

16. William Bunge, *Theoretical Geography* (Lund: Gleerup, 1966).

17. Waldo R. Tobler, "Geographical Area and Map Projections," *Geographical Review*, Vol. 53 (1963), 59-78.

18. Torsten Hägerstrand, "Migration and Area," in David Hannerberg, Torsten Hägerstrand and Bruno Odeving (eds.), *Migration in Sweden* (Lund: Gleerup, 1957).

19. Leslie Curry, "Quantitative Geography 1967," *Canadian Geographer*, Vol. 11 (1967), 265-279.

20. Michael J. Woldenberg, "Spatial Order in Fluvial Systems: Horton's Laws Derived from Mixed Hexagonal Hierarchies of Drainage Basin Areas," *Bulletin of the Geological Society of America*, Vol. 80 (1968), 97-112.

21. Gunnar Olsson and Stephen Gale, "Spatial Theory and Human Behavior," *Papers and Proceedings of the Regional Science Association*, Vol. 21 (1968); David Harvey, "Behavioral Postulates and the Construction of Theory in Human Geography," *Seminar Papers of the Department of Geography*, University of Bristol, Ser. A, No. 6 (1967).

22. Torsten Hägerstrand, *Innovations förloppet ur koro-logisk synpunkt* (Lund: Gleerup, 1953); Leslie Curry, "Central Places in the Random Economy," *Journal of Regional Science*, Vol. 7 (1967), 217-238; Leslie Curry, "A 'Classical' Approach to Central Place Dynamics," *Geographical Analysis*, Vol. 1 (1969); Julian Wolpert, "Migration as an Adjustment to Environmental Stress," *Journal of Social Issues*, Vol. 22 (1966), 92-102; Pred, *op. cit.*

23. Curry (1967), *op. cit.*

24. Hempel, *op. cit.*

25 Gunnar Olsson, "Central Place Systems, Spatial Interaction, and Stochastic Processes," *Papers and Proceedings of the Regional Science Association*, Vol. 18 (1967), 13-45.

26. Allan G. Wilson, "Notes on Some Concepts in Social Physics," *Working Paper from the Centre for Environmental Studies, London*, No. 4 (1968).

27. Walter Buckley, *Sociology and Modern Systems Theory* (Englewood Cliffs, N.J.: Prentice-Hall, 1967).

28. Kingsley Davis, "Population Policy: Will Current Programs Succeed?" *Science*, Vol. 158 (1967), 730-739.

29. Max Born, *Natural Philosophy of Cause and Chance* (New York: Dover, 1964).

30. William Kneale, "Scientific Revolution for Ever?" *British Journal for the Philosophy of Science*, Vol. 19 (1968), 27-42.

31. Hägerstrand (1953), *op. cit.*

32. Forrest R. Pitts, "MIFCAL and NONCEL: Two Computer Programs for the Generalization of the Hägerstrand Models to an Irregular Lattice," *Working Papers from the Social Science Research Institute, University of Hawaii*, No. 4 (1967).

33. Richard L. Morrill, "The Distribution of Migration Distances," *Papers and Proceedings of the Regional Science Association*, Vol. 11 (1963), 75-84.

34.   Duane F. Marble and John D. Nystuen, "An Approach to
the Direct Measurement of Community Mean Information Fields,"
Papers and Proceedings of the Regional Science Association,
Vol. 11 (1963), 99-109;   Richard L. Morrill and Forrest R.
Pitts, "Marriage, Migration, and the Mean Information Field:
A Study in Uniqueness and Generality," *Annals of the Association
of American Geographers,* Vol. 57 (1967), 401-422.

35.   David Harvey, "Geographic Processes and the Analysis
of Point Patterns: Testing Models of Diffusion by Quadrat
Sampling," *Transactions of the Institute of British Geographers,*
Vol. 40 (1966), 81-95;   Gunnar Olsson, "Lokaliseringsteori och
stokastiska processer," in Tor Fr. Rassmusen (ed.), *Regionale
Analysemetoder* (Oslo: Norsk Institutt for By og Regionforsk-
ning, 1967);   John Hudson, *Theoretical Settlement Geography*
(unpublished Ph.D dissertation, University of Iowa, 1967).

36.   Michael F. Dacey, "A County Seat Model for the Areal
Pattern of an Urban System," *Geographical Review,* Vol. 56
(1966), 527-542.

37.   William Feller, "On a General Class of Contagious
Distributions," *Annals of Mathematical Statistics,* Vol. 14
(1943), 389-400;   David Harvey, "Some Methodological Problems
in the Use of the Neyman Type A and the Negative Binomial Probability Distributions for the Analysis of Spatial Point Patterns," *Transactions of the Institute of British Geographers,*
Vol. 44 (1968), 85-95.

38.   Gunnar Olsson, "Complementary Models: A Study of
Colonization Maps," *Geografiska Annaler,* Vol. 50, Ser. B (1968),
1-18;   R.G. Swinburne, "Vagueness, Inexactness, and Impression,"
British Journal for the Philosophy of Science, Vol. 19 (1969),
281-299.

39.   Reginald G. Golledge, "Conceptualizing the Market
Decision Process," *Journal of Regional Science,* Vol. 7 (1967),
239-358;   Reginald G. Golledge, "The Geographical Relevance of
Some Learning Theories," *this volume* (1969);   Reginald G.
Golledge and Lawrence A. Brown, "Search, Learning, and the
Market Decision Process," *Geografiska Annaler,* Vol. 49, Ser. B
(1967), 116-124.

40.   Gerard Rushton, "The Scaling of Locational Preferences,"
this volume (1969).

41.   Peter R. Gould, "On Mental Maps," *Discussion Paper of the Michigan Inter-University Community of Mathematical Geographers,* No. 9 (1966);   Peter R. Gould, "Problems of Space Preference Measures and Relationships," *Geografiska Annaler,* Vol. 50, Ser. B (1968);   Peter R. Gould and R.R. White, "The Mental Maps of British School Leavers," *Regional Studies,* Vol. 2 (1968), 161-182;   Roger M. Downs, "Approaches to and Problems in the Measurement of Geographic Space Perception," *Seminar Papers of the Department of Geography,* University of Bristol, Ser. A No. 9 (1967).

CONCEPTUAL AND MEASUREMENT PROBLEMS IN THE

COGNITIVE-BEHAVIORAL APPROACH TO LOCATION THEORY

David Harvey

University of Bristol

In this paper I want to examine some of the problems that
arise from taking a cognitive-behavioral approach to location
theory.   The argument for this approach may be summarized as
follows.   Locational patterns in human geography are the physical expression of individual human decisions.   Locational
analysis must therefore incorporate some notions regarding
human decision making.   The simplest course is to set up
idealizations or to develop some descriptive device to summarize aggregate human behavior.   The idealization of rational
economic man leads us to the normative location models such as
those of Weber, von Thünen, Lösch, and their academic descen-
dants.   Empirical evidence suggests that it is possible to
conceptualize behavior as a stochastic decision process and
use probability distributions to discuss spatial behavior provided we are considering basically similar choices in a fairly
homogeneous population.   Descriptive mathematical functions
may then be used as the foundation for a stochastic location
theory -- a theory which has yet to be written and whose characteristics remain largely unknown in spite of recent general
formulations.[1]   But there are many situations in geography in

which these descriptive devices or idealizations are clearly
inappropriate and there is no alternative but to incorporate
very specific statements about the cognitive processes involved
in the act of decision.

We know that decisions are affected by attitudes, dispositions, preferences, and the like. We know, too, that mental processes may mediate the flow of information from the
environment in such a way that one individual perceives a
situation differently from another even though the external
stimuli are exactly the same. Each individual may be thought
of as making decisions with respect to his attitudes and in
the context of his perceptions. We also know that an individual's attitudes and dispositions may be affected, often
cumulatively over time, by the constant bombardment of stimuli
from the environment around him and by cultural conditioning.

We can perhaps abstract the sense of a cognitive behavioral location theory as follows. It should be able to handle
a process in which each individual decision maker, enclosed in
his own environment, reaches a decision which presumably maximizes some 'satisfaction' or 'preference' function defined
over his own dispositions and attitudes. To be a viable
theory it must also be able to handle problems of aggregation
that result either from any macro-locational analysis that is
required, or from needing to resolve the problem of conflicting
decisions in complex organizational structures. It would,
for example, be unthinkable to discuss organizational decisions,
such as those of government, without reference to the problems
of conflict resolution. In addition the theory will need to
handle the feedback effects from the environment to the decision maker. The simplest example of this kind of feedback is
the learning process which results when stimuli from the environment reinforce the attitudes and dispositions of the
decision maker.

At first sight the formulation of such a cognitive behavioral location theory appears an attractive but formidable task. It is attractive because it seeks to understand the decision process as it really is. I suspect that it also has intuitive appeal for many because it satisfies hopeful and hidden emotions about freedom of choice, individuality, and, ultimately, free will. In this respect it seems to function as a new and more sophisticated version of a dodo that refuses to die in geographic thought -- the notion that everything really *is* unique. The theory appears formidable because if it is ever to be anything more than a vague hopeful speculation, it will need to settle a whole host of conceptual and measurement problems in such a way that we can actually understand what has eluded the behavioral sciences as a whole -- viz, the *real* reasons why people behave with respect to their environment in the way that they do. It is unlikely, of course, that we will achieve complete understanding. The question therefore, is not whether we will construct a theory to explain everything about human decision making, but how far and how quickly we can progress along this road, constructing reasonable partial formulations as we go. It is useful to consider this question if only from the point of view of research strategy. If, as is possible, it will take enormous research effort and time to construct quite trivial behavioral formulations, then it might be advisable to shift our intellectual resources elsewhere. We thus find ourselves in the situation of the decision maker in the face of uncertainty. We cannot know the final answers until we have tried all possible strategies. But like all decision makers in the face of uncertainty we can generate certain expectations and make our decisions with respect to them.

It seems to me that we have three strategies open to us.[2] We can seek to extend 'classical' location theory with its

emphasis upon optimization techniques; we can seek to build
a stochastic location theory; or we can take a cognitive-
behavioral approach. These are not mutually exclusive or
easily separated strategies. They should be thought of as
three different focal points in the universal set defined by
the decision problem. It is useful to ask, however, what we
might expect the ultimate relationships to be between the
theories generated around these three foci. My own expectation is that these theories will have distinctive and only
partially overlapping domains. They will be complementary
rather than competitive. When policy issues regarding economic efficiency are involved normative economic location theory
cannot be replaced although there is, of course, plenty of
room for its improvement. There are undoubtedly many situations in which, either through the forces of competition or
through an inherent tendency among decision makers to seek out
optimal or close-to-optimal solutions, the normative economic
theory or extensions of it will provide us with a reasonable
and quite handy model for rather more complex decision processes. This is not, perhaps, a fashionable view among
geographers at present. It does, however, seem to me to be a
totally unwarranted inference that normative economic location
theory has no empirical relevance because we can so frequently
find deviations between patterns predicted from theory and
actual patterns. The trouble with normative location theory
is not that it is empirically irrelevant, but that we do not
know the circumstances in which it may be used as an empirical
device. It is uncontrollable rather than irrelevant in empirical work. Indeed, one of the side pay offs from formulating
an adequate cognitive-behavioral location theory may be to
improve our control over the empirical use of normative
models. I think it also reasonable to expect that stochastic
models of behavior will not be challenged by a cognitive-

behavioral theory in certain domains. Particularly when we
are concerned with the aggregate effects of countless individual
decisions (about which it is very difficult to collect any
definite information), it would seem senseless to attempt a
behavioral analysis of each individual decision and then try
to aggregate these up into a model that copes with the total
process. Probability theory, especially when given a relative
frequency interpretation, provides a set of extraordinarily
effective models for dealing with aggregate effects of decisions that are rather repetitive in form over time and space.
The general aggregative characteristics of migration, journey-
to-work, journey-to-shop, diffusion, and so on, can and will
most easily be handled by the use of stochastic models. Considering the well-documented statistical regularities which
can be observed in human behavior in space, it appears very
reasonable to expect the integration of many of these concepts
into some basic stochastic location theory. We are then left
with a 'residual' domain of events which cannot be handled by
the normative or stochastic location theories.

This domain will presumably be colonized by a cognitive-
behavioral location theory. Let us consider the kind of
event which this domain might contain. There may, for
example, be policy problems in which criteria of economic
efficiency are either difficult to define or irrelevant, or
situations in which the goals of societies, groups, or individuals are clearly noneconomic. A cognitive-behavioral
theory will need to handle the complex problems of value
judgements, utility scales, and so on. There may also be
situations in which a very few decision makers have a dispro-
portionate effect upon spatial patterns. At the micro-level
we may be concerned with how an individual or a small sample
of individuals is acting in a given situation (for example, a
dozen or so farmers in a particular area). In this case

as the sample size increases so the stochastic models may become more relevant. At the macro-level we may be concerned
with the decisions of a few individuals in government or big
business since these decisions can have an enormous effect upon
regional development. In these circumstances it is vital to
understand the attitudes and dispositions (particularly political dispositions) of those involved in the decision process,
and to understand how alternatives are perceived, searched, selected, and implemented. The domain of a cognitive-behavioral
location theory will thus range from cross-cultural variation
in value judgements and perceptions through individual choice
behavior to group decision making processes.

It is pertinent to speculate now about the prospects for
formulating reasonable theories in the cognitive-behavioral
domain. Again, we are forced to speculate in the face of
uncertainty, but certain expectations can be generated to help
make decisions on research strategy. I rest these expectations on my knowledge (which is obviously incomplete) of the
conceptual and measurement apparatus currently available to
us. Future research results will undoubtedly alter the pic-
ture considerably, but since these are unknown it seems best
to base our assessment on current information. It is useful
to concentrate on conceptual and measurement apparatus because
provision of these in a *sine qua non* for successful theoretical
formulations and their application.

Concepts provide us with analytic power, while measurement procedures provide us with the necessary techniques to
pin down our analyses to the world of experience. Concepts
and measurement techniques are not independent of each other.
Strict operationalists would claim, for example, that every
measurement procedure must logically bear a unique relationship to concept. They would also argue that each concept
can only be defined by reference to the procedures which are

used to gain knowledge of it -- and these procedures often involve measurement. I do not wish to take such a strict operationalist view in this essay. The link between concepts and measurement procedures is undeniable, but we can usefully separate them and on occasion regard them as being quite different from each other.

A. Concepts in General

I take it as axiomatic that we can hope to handle the complex world of behavior only by formulating firm concepts with generally agreed meanings. If we are to communicate our ideas we must first agree upon some way of assigning meaning to terms and this amounts to agreeing upon some procedure of definition. To do this we must define concepts and relate them to experience.

It is worth distinguishing between *theoretical* concepts and *empirical* concepts. This distinction is important because the assignment of meaning differs radically according to type, and because it is important not to mix the two. Theoretical concepts can be defined implicitly. Empirical concepts can be defined explicitly. To understand this better we need to define our own terms and we can best begin by asking what we mean by the term theory. I shall simply regard it as an abstract calculus in which the terms can be defined by their syntactical function rather than by reference to their empirical interpretation. This is perhaps a difficult idea to grasp, but it can be illustrated most easily by mathematical examples. The terms point and set in mathematical formulations have no empirical meaning, and if we wish to give them meaning we can do so by examining the way they function in the calculus. It is rather like defining the rook in chess by specifying the field of play and the rules governing its movement. Now it is possible to relate these theoretical concepts to perceptual experience by establishing a set of

'correspondence rules' or 'epistemic correlations'[3] between
them and empirical concepts. A mathematical point may thus
be correlated with a dot on a piece of paper. This epistemic
correlation does not, however, exhaust the possible semantic
interpretations of the theoretical concept. In the applications of probability theory the same concept of a point, this
time located in a sample space, may be interpreted as an outcome of an experiment. Establishing epistemic correlations
provides a rather different way of assigning meaning to a
theoretical concept. But in this case the assignment is
implicit and incomplete. Now it was the operationalist argument that the only worthwhile manner of assigning meaning was
by way of such epistemic correlations. This is now generally
regarded as being unnecessarily restrictive and not very advisable because it makes the theoretical structure incapable
of further application and further growth.[4]

Empirical concepts are capable of explicit definition with
respect to experience. We can provide ostensive definitions
or provide operational definitions. It is also a characteristic of empirical concepts that when we give them lexical definitions, we can replace the term being defined by the *definitions* without any loss of information. Thus we can exhaust
the meaning of an empirical concept if we so wish by specifying fully the operations by which knowledge of that concept is
obtained.

We possess two languages. The language of theory contains concepts which can be defined syntactically although
they can also be interpreted in terms of real world experience
via epistemic correlations. The language of empirical investigation contains concepts which can be defined operationally,
explicitly, and completely. If theory is to be of any use we
need to be able to translate from one language to the other.
This is not always easy to do in the social sciences.[5]

Empiricists complain that social science theory is in general incapable of empirical interpretation and therefore not worth bothering with. Theoreticians complain that their theories remain uninterpreted because empiricists have failed to establish the kinds of concept which will allow epistemic correlations to be made. There are, of course, many situations in which this translation has successfully been made. In a stochastic location theory, for example, the empirical concept of a town (which can be given an operational definition) can be translated into a mathematical point in a sample space. In most cases these translations run from empirical concepts to mathematics rather than from empirical concepts to social science theory. This latter form of translation is not helped by the fact that we often use the same term to represent both theoretical and empirical concepts. A point, for example, has meaning in both language systems and it is important to differentiate in which sense it is being used. It might almost be worthwhile to adopt a notational system to make explicit which language we are using when there is any possibility of ambiguity.

Let us now consider some of the concepts available to us in behavioral science in the light of these two kinds of language. There is certainly no shortage of theoretical concepts. Indeed, looking round the behavioral sciences it is difficult to know where to begin. It is easiest to proceed by example. I shall therefore begin by examining two concepts of considerable importance to our argument, *economic rationality* and *satisficing behavior*. The first lies at the very center of economic location theory and the second, judging by its frequency of use in geographic literature, is an important if not central concept in the cognitive-behavioral approach to location theory.

The concept of economic rationality has been an extraordinarily fruitful one. It functions as one of the primitive
terms of economic location theory and, like most primitive
terms, it can be used to generate derived terms -- concepts of
marginal behavior, pure and perfect competition, profit maximization, and so on, readily spring to mind. From this it is
easy to see that the concept of economic rationality can be
given a firm syntactical definition in terms of its function
in theoretical economics which includes, of course, the economic theory of location. The concept of satisficing behavior
has a much shorter history and it would be churlish, therefore,
to expect it to be as well developed syntactically as the concept of economic rationality.   But the fact remains that it
has not been a very fruitful concept.   In some respects it is
theoretically ambiguous.   I think the basic trouble with it
is that it is a negative rather than a positive concept.
It is designed to explain the empirical shortcomings of economic theory rather than to generate theory in its own right.
But it is still useful to ask how it might function in a
cognitive-behavioral location theory.

The concept of satisficing behavior is a very confused
one.   It has several connotations.   Let us consider, for
example, the relationship between satisficing behavior and
optimizing behavior.   We can, if we wish, regard satisficing
behavior as a form of optimizing behavior in which the criteria used are non-economic.   Instead of writing an objective
function which maximizes profits, for example, we might write
one that maximizes leisure time subject to the constraint that
a certain minimum income is achieved.   There is another possibility.   Satisficing behavior may be regarded as optimizing
behavior (of any kind) with respect to a number of pre-selected
alternatives out of a much larger (sometimes infinite) set of
alternatives.   In this case the concept of satisfaction may

refer to the decision-maker's intuitive assessment of the
adequacy of his pre-selection process.   I suspect that this is
what Simon actually meant by the term 'bounded rationality'.[6]
But there is yet another possible interpretation.   Even given
a bounded choice and noneconomic criteria, the decision
maker does not seek *any* optimal solution.   In this case satisficing behavior means non-optimizing behavior.   I suspect this
last interpretation is theoretically barren, yet it is an interpretation that geographers are rather partial to.   It is theoretically barren because theories are deductive structures and
this interpretation of the concept of satisfaction is only
capable of being exploited inductively.   In short, if we
accept this interpretation almost *any* form of decision-making
behavior could follow from it.   Now it may be the case that
decision making really is characterized by nonoptimization in
the world of experience -- I do not wish to deny this possibility -- but a theoretical concept of satisficing behavior
is worse than useless if it merely refers to this possibility
without giving us any further clue as to how we might handle
such situations.   My conclusion is that we either need to
interpret satisficing behavior as some form of optimizing
behavior, or we must abandon the concept and seek for theoretical concepts which do give us some control over the analysis
of nonoptimizing behavior.   We will consider certain possibilities from perception studies later.

How can these two concepts of economic rationality and
satisficing behavior be given an interpretation in empirical
language?   What kinds of epistemic correlations can be
established?   In neither case, of course, can these epistemic
correlations exhaust the theoretical meaning of the concept,
and several epistemic correlations are possible in each case.
Again, it seems to me that the concept of economic rationality
has been given more and more fruitful empirical interpretations

than has the concept of satisficing behavior.   There are
many empirical situations in which it is possible to define
profit maximizing or cost minimizing behavior in an operational sense and most of the techniques of operations research
are available for discussing these problems empirically.
Operational definitions can be found for concepts of profit,
loss, cost, and so on.   There are many situations in which it
is possible to measure exactly and thereby achieve operational
control over the concepts concerned.   In other cases operational definitions are less easy to come by.   The theoretical
concept of profit, for example, cannot always be given an
empirical interpretation.   Do we mean short-term or long-term
profit and how long is the long term?   How do we measure indirect nonmonetary benefits?   What does profit mean in the
context of the social system as a whole?   So the same concept
can be given an empirical interpretation in some situations and
not in others.   I suppose the main objections to the concept
of economic rationality (and its derivatives) arise from the
failure to discriminate between these two different kinds of
situation.   But to dismiss the concept because it has been
grossly misused is to throw the baby out with the bathwater.
If we wish to control the use of the concept of economic
rationality in an empirical context we can do so by a careful
appraisal of the measurement procedures used to quantify it.

How can the concept of satisfaction be given an empirical
meaning?   How can we measure it?   The answer depends, of
course, upon what theoretical interpretations we are giving to
the concept.   If we regard it as nonoptimal behavior (of
any kind) then there is only one way we can operationalize it
in its present form.   We can generate expectations between
expected and observed behavior as some measure of the degree
of satisficing behavior.   Now there seems to be something
fundamentally unsatisfactory about this procedure.   It

assumes the adequacy of the economic optimizing model in the
measurement of satisfaction! Deviations between expected and
observed behavior may, of course, be explained in a number of
different ways. There may be errors of specification in the
optimizing model, errors in estimating the parameters, and so
on. The deviations are of inherent interest, but there is no
guarantee (short of perfection in the optimizing model and in
its calibration) that they are realistic measures of satisficing behavior. If, on the other hand, we regard satisficing as a form of optimizing behavior, then it may be possible
to operationalize the concept in the same way that we can
operationalize the economic model. Consider the various ways
in which we might specify an objective function in an operational manner. We may maximize leisure time, minimize labor
input, minimize effort, maximize security levels, and so on.
These can be operationalized. We cannot maximize happiness,
joy, or, for that matter, satisfaction, in any operational
sense. All we can do, therefore, is to translate happiness or
satisfaction into empirical concepts such as leisure time,
security levels, and so on. As always, the translation is incomplete. But unless we are willing to perform such translations we cannot hope to discuss the concept of satisfaction
in empirical terms.

It is worth noticing, however, that theoretical concepts
which started out by being polar opposites are now operationalized so that they are different in degree not kind. Both
economic rationality and satisficing behavior are interpreted
as optimization behavior, and the only difference between them
is the nature of the objective function -- economic rationality
presumes profit is to be maximized, satisficing behavior presumes it is some other quantity, such as leisure time. It
could be argued, however, that there is a real difference in
the nature of the constraint sets. If we state, for example,

that a certain minimum income level must be achieved while maximizing leisure time, then the concept of satisfaction enters into the problem by way of the arbitrary choice of this minimum level. Some decision makers might set it at $10,000 and others at $20,000, and so on. Without this information the model would certainly not be realistic. But this particular arbitrary decision does not strike me as being any different in principle from the kind of arbitrary decision necessary in the economic model regarding the resource constraints (how much labor to use, and so on). It is also worth noting that in neither case is the concept of omniscient understanding employed. In empirical work the concept of profit maximization has never assumed total knowledge. The alternatives evaluated are usually a few out of all possible alternatives. In both cases, therefore, we make a choice from a bounded set. Therefore, if we accept the idea of satisficing behavior as optimizing behavior, the difference between the former and economic rationality narrows very appreciably in the language of empirical research. There is also a considerable pay-off to be had from studying theoretically and empirically the relationships between solutions generated by different objective functions. This is not simply an interesting question -- it is a vital one. Many of the social problems we currently face can be conceptualized in these terms. Pollution problems might be considered in terms of a conflict between optimization over the total social system and optimal behavior on the part of individuals operating within that system. Conservation problems might be regarded as a conflict between profit maximizing and, say, leisure maximizing. I have not space, however, to launch into a discussion of these kinds of question here.

If satisficing behavior is regarded as nonoptimizing behavior then the above argument is unacceptable. The concept becomes a negative one which merely serves to remind us, often quite appropriately, that profit maximization is not everything in life.   But the concept is theoretically and empirically barren.   It must, therefore, be replaced with positive concepts which bear theoretical fruit and, preferably, which can be operationalized in empirical work.   There are plenty of such concepts available to us.   I shall therefore again proceed by example and take a more detailed look at the concepts and measurement procedures available for the study of perception.   The concept of satisfaction is often associated with the idea that the decision maker proceeds on the basis of his images or perceptions and as we saw at the beginning of this essay, the notion of a difference between the 'perceived world' and the 'real world' is in any case important in the cognitive-behavioral approach to location theory.

B. Theoretical Concepts in Perception Studies

The concept of perception is itself rich in ambiguity.   In some cases the concept is used "to designate a world view, an outlook on life, or some other very general cognitive product."[7] This interpretation has little to recommend it from an analytic point of view.   It does remind us, however, that there are some extraordinarily interesting problems in phenomenological philosophy and that 'images' of great generality may have tremendous significance to human decision making.[8]   But in this very general sense the concept of perception is no better than the general concept of satisfaction.   If we are to handle problems we need more precise definitions than this.   Even in technical terms, however, the concept has been given quite different interpretations and it is difficult to avoid the conclusion that it often refers to several different processes. Much work is concerned with the physiological aspects of perception.   The processes are essentially visual, auditory, etc., and the emphasis of study is upon those variables that impinge

directly upon the senses.   This kind of approach is only
indirectly relevant to the construction of a cognitive-behavioral location theory.   We are more interested in the social
aspects of perception.   Not, of course, that the two are independent of each other.   But the focus of interest is different.   Within the social perception field, however, there
are several competing frameworks for examining perceptual processes.   Geographers have, at various times, been attracted
to different formulations.   Some refer to *gestalt* perception
studies,[9] some to field-theory frameworks,[10] and so on.[11]
It is hardly surprising to find, therefore, that no generally
agreed definition of perception can be supplied.   *Precise*
definition is probably neither necessary nor desirable, and
Warr and Knapper regard overemphasis upon precise definition
as a danger to be avoided: "it is far from disconcerting that
books on perception cannot open with a complete definition of
this concept..... it is clear that in some sense perception
involves an interaction or transaction between an individual
and his environment; he receives information from the external world which in some way modifies his experience and behavior.   But beyond statements of this order of generality
there are few formulations which are universally accepted.
Writers with different backgrounds and objectives tend to
emphasize different aspects of the process, so that various
approaches are reflected in varying definitions."[12]

It is possible therefore, to choose our own definition of
perception (within limits) according to our objective.   In
our case the objective is to formulate a cognitive-behavioral
location theory.   It seems reasonable to conclude that the
definition we choose for the concept of perception should be
one relevant for the discussion of spatial behavior.   Above
all, we would prefer to be able to predict spatial behavior
on the basis of our understanding of perception.

At this point it is worth introducing a distinction
between attitudes and perceptions. Warr and Knapper[13] suggest that attitudes are relatively permanent structures which
hold in the absence of any particular stimulus, whereas perceptions are more flexible and transitory and only occur in
the presence of a stimulus. Obviously, there are strong
interactions between attitudes and perceptions defined in the
above manner -- attitudes are presumably formed by perceptual
experience and, in turn, affect the 'perceptual readiness'
of the individual. I think the difference between attitudes
and perceptions is particularly useful to us. If we accept
the definition of an attitude as a 'learned predisposition to
respond to any object in a consistently favorable or unfavor-
able way,' then we ourselves would, I think, be predisposed
to think that attitudes have a powerful influence over spatial
behavior. But recent work in fact suggests that a person's
attitude to an object is *not* a major determinant of his behavior with respect to that object.[14] There is some effect,
of course, but this is by no means as strong as usually is
assumed. If this is the case, then we can afford largely to
ignore the problem of attitudes in seeking for a cognitive-
behavioral location theory. Is it possible then, to make the
concept of perception a corner-stone for such a theory?
There are several reasons why I think the answer to this question should be positive. We know, for example, that many
decisions are 'impulse' decisions -- decisions about purchases
of goods, migration, investment, and so on, are frequently of
this sort. A concept of perception which refers to transitory events under direct stimulus seems much more useful for
the analysis of these kinds of decisions than does the concept
of an attitude. It would be difficult to prove what proportion of decisons are of an impulse type (rather more than we
usually cater for I suspect), but even in those cases where

attitudes are important, it is quite possible that they are so simply through their effect upon perception. In other words, perception as we are here defining it, might be regarded as the central node in a network which brings together cognitive processes and environmental stimuli and which projects to the act of decision. I write 'might' advisedly because if we are to gain anything from this broad conceptualization, we must develop a firmer theoretical framework for analysis and establish the necessary epistemic correlations to facilitate empirical work. Because it seems to me to be important to avoid too great a gap between the language of theory and research, I shall endeavour to clarify some theoretical problems by taking a close look at empirical concepts of perception and in particular at the measurement procedures which may be used to gain knowledge of them.

C. Empirical Concepts and the Measurement of Perception

Ogden and Richards long ago remarked that "perception can only be treated scientifically when its character as a sign-situation is analysed."[15] I want to tackle the problem of measuring perception via semiotics (the theory of signs) partly because this provides a coherent framework for examining the measurement problem, but also because it provides us with a convenient way of bridging the gap between theory and empirical research. There is no need to define a sign with any precision -- it may be a symbol (such as a word), a photograph, a map, an object, an experience, and so on. Morris,[16] who has pioneered the theory of semiotics, prefers to leave the meaning open and considers the nature of signs in terms of a general sign-process. From our point of view the importance of Morris's presentation is the way in which this sign-process relates to behavior.

Morris regards a sign-process as a five point relation in which a *sign* sets up in an *interpreter* (in our case the

decision maker) the disposition to act in a certain kind of way
(called the *interpretant*), to an object or event (called the
signification) in the *context* in which the sign occurs.
Morris goes on to suggest that signification is tridimensional
and that it can be correlated with perceptual, manipulatory,
and consummatory aspects of action:

"The organism must perceive the relevant features of the
environment in which it is to act; it must behave toward
these objects in a way relevant to the satisfaction of its
impulse; and if all goes well, it then attains the phase of
activity which is the consummation of the act..... A sign is
designative insofar as it signifies *observable* properties of
the environment or of the actor, it is *appraisive* insofar as
it signifies the consummatory properties of some object or
situation, and it is *prescriptive* insofar as it signifies how
the object or situation is to be reacted to so as to satisfy
the governing impulse."

At this juncture it is worth comparing the terminology
used by Morris to describe the three aspects of signification
with the terminology developed to handle the various components of psycholinguistics and perception.[17] The designative
aspects of signification may thus be regarded as essentially
similar to what are called denotative meanings or the attributive component in perception.[18] Perception, it has been
remarked, invariably involves an act of categorization.[19]
A sign or stimulus is thus placed in a certain class by virtue
of its defining attributes. When we ask what a sign desig-
nates, therefore, we ask what category it belongs to. This
raises some fascinating problems regarding the interaction
between perception and language since it is the latter which
mainly determines the categories into which stimuli may be put.
This is not the place, however, to debate the pros and cons
of the Whorfian hypothesis that cognitive behavior is

influenced by the semantic structure of language, although
this hypothesis, if true, has important implications for the
analysis of cross-cultural differences in perception.[20]   The
appraisive aspect of signification may similarly be related to
connotative meaning of the *affective* component of perception.
A sign may provoke in us feelings of attraction or repulsion,
like or dislike. From our point of view this is an important
aspect of perception, because if a sign provokes a positive
attraction then we can anticipate positive behavior with respect to it; if not we can anticipate negative behavior. If
we can find some method of measuring the effective component
of perception we will have gone some way to measuring a component of satisficing behavior. The *prescriptive* aspect of
signification may also be related to the *expectancy* component
of perception, although the relationship is not a perfect one
by any means. A sign may provoke in us certain expectations
on the basis of which we may make predictions. Each sign
therefore stands in some relation to other signs and on the
basis of these relationships we may make inferences of one
sort or another. If, for example, the sign is the word combination 'rich suburb' then we generate certain expectations
about the kinds of houses there, the people who live in them,
and so on. If we wish to act in some way with respect to a
'rich suburb' we also generate certain expectations about what
is a feasible course of action (a planner may find it to be
the optimal location for a sewage processing plant but he may
infer immediately that such a plant will be impossible in a
'rich suburb'). If we are to understand this component of
the sign process, then we must understand the structure of
associations which the individual possesses. Since these
structures are relatively permanent it seems best to conceptualize them as attitudes rather than as perceptions. I
suspect that it is at this point in the sign-process that

attitudes are most important and we cannot, therefore, ignore attitudes altogether in our analysis of the perceptual process.

Some signs, such as the word 'man', are primarily designative (attributive or denotative), some, such as the word 'bad', are primarily appraisive (connotative, affective), while others, such as the word 'should', are primarily prescriptive. But all signs have some signification on all three dimensions.

How does this general notion of a sign-process relate to the measurement of people's perceptions? The necessary operational concepts for the measurement of human perceptual behavior are provided by stimulus-response psychology.[21] The measurement of perception involves scaling (on some appropriate model) the responses of individuals to some stimulus. I do not wish to consider the problem of what constitutes an adequate scaling model and I shall therefore take it for granted that the problem of matching the responses with some scaling system can readily be overcome. I want to concentrate upon the nature of the stimuli. All stimuli may be regarded as signs (although not all signs function as stimuli, e.g. certain logical signs such as 'and'). This elementary fact provides the link between the generality of the sign-process and the particularity of measurement procedures in perception studies. The response which we scale is a disposition to act in a certain way as a result of an interpreter experiencing the sign. The problem of measurement in perception may thus be regarded as a question of how we select signs and how we can control the sign-process so as to yield meaningful insights into spatial behavior.

It is useful at this stage to divide signs into *symbols* and *signals*. A symbol is a substitute sign, signifying what the sign for which it is a substitute signifies. If a sign is not a symbol we will call it a signal. Generally speaking our direct experience of an environment may be regarded as a sign-process in which the signs are signals. The architectural form of a street, the morphology of a landscape, and so on, are mainly perceived through signals. Some elements of this experience may be conceptualized as symbols in certain contexts (e.g. the form of a church, or the layout of a village). We may, however, read about the same things in a book, in which case all the signs are symbols which are presumably chosen to represent the salient signals experienced in the environment. Now the relationship between signals and symbols is an important one. In general our measures of perception in the social sphere are dependent upon symbols rather than signals. It is too expensive to take people to an area and monitor their responses during their stay there, and we therefore administer the symbol 'Devon' and measure their reactions to that. Most investigations into geographical perception are thus going to be through measuring reactions to symbols such as words, photographs, pictures, maps, and so on. Yet actual behavior is often going to be determined by the perception of signals (shop window displays, attractive views, and so on). In these situations we need, therefore, to calibrate the relationship between signal and symbol.

Let us consider the ways in which the signal sign-process and the symbol sign-process must necessarily differ. The context is obviously different. Symbols may be reacted to in a warm comfortable room, signals are experiences in a totally different environmental situation. The symbolic process can only indicate certain selected features of an object or generalize about it in such a way that a lot of information is lost (e.g. the relationship between the signals that emanate from Devon and the symbol 'Devon'). The signal process shows an object in its total environment. This is not necessarily a disadvantage for we have far greater control over the

symbolic process as a stimulus. We can cut out background
information on a photograph of a house, but we cannot take
somebody to a house and ask them to look at it as if the background did not exist.

In defining a symbol it was suggested that it should
signify the same thing as the signal it is designed to represent. Yet symbol and signal cannot be the same in every
respect. We therefore need some way of defining the equivalence between signs. We cannot do this independent of a motivational situation.[22] The response to the symbol 'Devon' for
example, will depend upon whether we are contemplating taking
a holiday or a permanent job. We can only talk of equivalence in the same motivational context. Equivalence cannot
be established either without specifying the particular dimension of signification on which signal and symbol are to be
equivalent. We might describe a suburb as 'nigger', 'negro'
'black' or 'colored' and on the designative dimension these
symbols are not far apart, but on the appraisive dimension the
symbols are very different. We may thus speak of signs which
are designatively equivalent (they refer to the same object),
appraisively equivalent (they arouse similar emotions of attraction and repulsion in us) and prescriptively equivalent
(they suggest similar lines of action).

What of the symbols themselves? Since most human communication is through symbols and the learning process is
largely an acquisition of symbolic tools, we must expect that
symbolic representations are important in their own right and
in many cases they attain a meaning which is independent of
the signals which they were initially designed to represent.
Consider the symbol 'Devon'. The relationship between it and
the signals that emanate from that area of land called Devon
is obscure. Suppose we want to predict decisions to migrate
to Devon. Is that decision made with respect to the signals

that emanate from Devon or is it made with respect to the symbol 'Devon'? Symbols are often more important than signals in determining behavior, for the symbol may itself have an appraisive signification of some sort which has nothing to do with the signal process. Herein lies the difference between an individual's image of something (drawn from its symbolic representation) and the reality of that thing. Many an emigrant has been attracted to America, the land of opportunity, only to find that the image has been misleading in certain respects.

This raises an intriguing problem for perception studies in geography. We can, if we so wish, try to measure the designative, appraisive, or prescriptive aspects of the symbols themselves. We may study people's reactions to words such as 'Devon' or 'Vermont' and try to establish the image attached to that symbol. We may also examine the perception of maps, photographs, and so on, in their own right. On the other hand we may use the same symbols to try and study an individual's preferences with respect to a set of signals, such as the physical attractiveness of Devon or Vermont, or the signals represented by a map or photograph. Clearly, there is the possibility of confusion in the interpretation of measures obtained by using symbols as stimuli. One individual may interpret a symbol stimulus entirely at the symbolic level, another may do so with respect to the signals which a symbol represents, and some may have a mixed interpretation (partly symbolic and partly in terms of the signals). Which interpretation an individual gives to the symbolic stimulus will depend on a number of factors, but the one which is of obvious importance will be the amount of information and direct experience which a person has of the symbol and the signals which it represents. If a symbol such as a word or a map represents an area that I know well, then I am likely to

associate the signals emanating from that area with the symbol,
but if it refers to an area I have never had any experience
of, then I will interpret it in symbolic terms. This symbolic
interpretation will depend on the contexts in which I have experienced the symbol before. Suppose the only times in which
I have come across the symbol 'Devon' are on travel posters
which beçkon me to holiday in beautiful Devon.   Then it is
likely that I will react favorably to the term on the appraisive dimension. If I have never come across a symbol before
-- say, 'Clackmannanshire' -- then I have no word associations
upon which to base my judgement.   In this case I might simply
judge the symbol on how nice it sounds.

Now it is quite usual in perception studies in geography
to attempt to compare measures on different symbols over different people.   This is clearly a fairly 'noisy' procedure
if the above analysis is anything more than nit-picking.   Let
me give two examples.   Suppose we ask people to rank counties
or towns in order of their preference for living in them.   We
are here providing a list of symbols and measuring people's
responses to them in a given motivational setting.   In general
we may expect that people have experience of towns or counties
close to them (given the usual distance effect upon spatial
behavior).   We may therefore expect that the symbols of counties or towns close by will tend to be interpreted at the signal level.   Further away the symbolic interpretation is
likely to be more important.   I wonder if the tendency for
people to react favorably to their home area and then to comment favorably upon certain other areas further away,[23] can be
explained by the mixing of a local signal effect with a more
general symbolic effect?   The same sort of comment can be
made about the use of maps in geographical perception studies.
If the map is used as a stimulus, do the results refer to the
perception of the land represented by the map or to the perception of the map itself?   Is it the country represented by

the map which looks 'interesting' or is it the symbolic form
that is 'interesting'?   There are additional difficulties in
using the map as a stimulus since it is a particular type of
symbolic representation over which people have unequal command
-- some are good map readers and some cannot understand the
relationship between the map and reality at all.   Measures
derived from maps as stimuli may, therefore, reflect the varying ability to think in terms of abstract spatial schema rather
than measures of preference for those areas which the map
represents.

In both of these examples we are really scaling several
different things simultaneously -- the ability to read the language in the particular symbolic form we have chosen, the response to the symbols themselves, and the response to the signals which the symbols represent.   The variation in our
measures may thus be explained by reference to several different sources of variation.   It could be argued, of course, that
variation from unwanted sources may be regarded as random noise
or that in any case it is the totality of the measure which is
important if we are concerned with predicting geographical
behavior.   These arguments may be correct.   But it seems important to have some empirical evidence and this would not be
too difficult to collect.   We could, for example, compare a
population group which has physical experience of Devon (and
which presumably builds images with respect to the signals)
and compare their images with those of population groups which
lack such indirect experience.   There are some interesting
experimental possibilities here.[24]

The theory of signs or semiotics as it is usually called
provides a useful framework for the study of perception in
geography.   At the theoretical level we can conceptualize
geographic behavior as the result of some sign-process in
which individuals are reacting to signals and symbols.

Semiotics also has empirical relevance for it provides us with
a framework for understanding the measurement problem and in
particular the concepts of a signal and a symbol provide us
with a means for differentiating between signs and establishing
some sort of control over the stimuli used in measurement.
The tri-dimensional concept of signification also has both
theoretical and empirical relevance. It provides an underlying model for understanding the various facets of the perception process. We can, if we wish, examine sign-processes
on one of the dimensions only. Most attention has been paid
to the appraisive dimension and there are techniques for
measuring the performance of a sign on this basic dimension --
the most important being the semantic differential which was
specifically developed for the study of the connotative aspects of meaning.[25] Theoretically we may distinguish between
reactions based on signal-processes and behavior based on
symbols. I suspect that we may find, for example, that local
shopping behavior can best be understood as a signal-process,
whereas long-distance shopping to large centers may be conceptualized as a symbol-process. Other aspects of behavior in
the city may be affected by signals -- Lynch, for example, has
shown how distinctive architectural features in a city act as
signals and from this we may expect that individual travel
behavior may be analyzable by a study of the signals on various
city routes.[26] On the other hand processes such as emigra-
tion, migration over long distances, and so on, are more probably affected by the images attached to symbols. I suspect
in many cases that signal and symbol components will be inter-
mixed in various proportions.

There are many aspects of the sign-process which I have
not considered of course. The human mind has a limited
channel capacity and cannot accept all the signals which an
environment sends out. The process of selection of signals

is therefore important. We can think of signals of varying strength -- a cathedral sends out a massive signal whereas an ordinary house in an ordinary suburban setting sends out a very mild one (unless it happens to be home). We can measure the strength of various architectural symbols (Lynch's work is again interesting in this respect) and thereby make estimates of the probability that an individual will or will not perceive a particular signal. I have not space, however, to consider these various other aspects of perception in a geographical setting.

D. A Cognitive-Behavioral Location Theory?

I now want to draw upon the preceding discussion of theoretical and empirical problems to try and assess the prospects for formulating a cognitive-behavioral location theory.

It is inevitable in the early stages of any investigation that the concepts (both theoretical and empirical) which we use will be loosely defined. To demand exactness, precision and rigor in the use of terms in the initial stages of investigation can only result in 'the premature closure of our ideas' and hence have a pernicious effect upon the direction of research.[27] Yet we cannot afford to take this as a charter for interminable vagueness in our concepts. Indeed the degree to which we succeed in reducing this vagueness is a measure of our progress. But the reduction should be real and not spurious and forced. At the present time the concepts available to us for formulating a cognitive-behavioral location theory are rich in ambiguity and I believe this to be necessarily so. I cannot avoid the suspicion, however, that the concepts current in geographical writing are unnecessarily vague. The concept of satisficing behavior is an excellent example. Similarly, we could do much more to pin down a conceptual apparatus for the study of perception in geography than we have done. The trouble here, of course, is that we must

necessarily rely upon perception concepts as they are formulated in psychology and here we have considerable freedom of
choice. Given our concern with geographical behavior, we are
likely to find that much of the psychological literature on
perception (and the conceptual apparatus contained therein)
will be irrelevant to our purpose. But without a strong command of the psychological literature (which I for one do not
possess) it is difficult to determine which presentations are
useful to us and which are irrelevant. I am forced to conclude that we cannot hope to formulate a cognitive-behavioral
location theory unless we breed geographers who have a strong
command over the literature of behavioral science in general,
or find behavioral scientists who are interested in geographical problems. The same sort of comment, however, can be made
with respect to economic and stochastic location theory. In
spite of the lip-service which geographers have paid to the
former, there are few of us capable of adding to that theory
mainly because we do not know enough economics. A stochastic
location theory will similarly demand an adequate command over
the mathematics of probability theory. But both economic and
stochastic approaches to location theory have a distinctive
advantage over the cognitive-behavioral approach because the
concepts involved are far less ambiguous and much more easily
pinned down. I therefore suspect that we may get further more
quickly in developing economic and stochastic theory than we
will in developing the cognitive-behavioral theory. If we
are searching for immediate pay offs, therefore, we will do
better to invest our time in furthering normative economic
theory and in formulating stochastic theory -- I believe the
second possesses the greater untapped potential. But this
has the unfortunate effect of leaving the domain of cognitive-
behavioral events empty of any kind of formulation at all. I
am sure that this is unacceptable and undesirable.

It is undesirable because the cognitive-behavioral domain contains problems which are real enough and significant enough -- we surely cannot afford to ignore them.   I suspect it is unacceptable because this domain is of enormous intrinsic interest.   The problem of perception, for example, is so basic to everything we do and think and it is so basic to our understanding of knowledge itself.   The cognitive-behavioral domain undoubtedly poses the greatest challenge of all.   Stochastic theory, although of enormous potential, avoids so many intrinsically interesting problems.   It may give us excellent predictive control over aggregate events but I fear it will never yield us a really deep understanding of process.   I also take the view that research should not only be useful -- it should be stimulating and fun.   Here the cognitive-behavioral domain has distinct advantages!

We can enter this domain however, with the expectation of obtaining only very limited results.   I would doubt if anything very satisfactory will emerge in the way of general theory until the year 2,000 A.D. or so.   But certain small problems will, I think, prove tractable to theoretical analysis and empirical investigation.   On the theoretical side, Isard and Dacey[28] have shown the way.   On the empirical some of the studies on the perception of resources and place preferences[29] have indicated that interesting and useful results can be obtained.   If we refashion and sharpen our conceptual tools and improve our understanding of the measurement process, I have no doubt that substantially better results can be obtained. The one thing we cannot afford, however, is to indulge in that particular kind of intellectual laziness which regards it as unnecessary and foolish to try to eliminate vagueness and ambiguity in our conceptual apparatus.   In this we must remain perpetually aware of the trade-off that exists between unnecessary ambiguity and premature rigor and adapt our research strategy accordingly.

NOTES

1. See, for example, A.G. Wilson, *The Concept of Entropy in Urban Modelling* (London: Pion, forthcoming).

2. I have discussed the background to these strategies in D. Harvey, "Behavioral Postulates and the Construction of Theory in Human Geography," *Bristol Seminar Paper*, Series A, No. 6, (1967), (to be published in *Geographica Polonica*, 1969).

3. E. Nagel, *The Structure of Science* (New York: Harcourt, Brace, and World, 1961), and F.S.C. Northrop, *The Logic of the Sciences and the Humanities* (New York: Macmillan, 1947).

4. See R.B. Braithwaite, *Scientific Explanation* (New York: Dover, 1960), 77.

5. H.M. Blalock and A. Blalock, *Methodology in Social Research* (New York: McGraw-Hill, 1968), 5-27.

6. H.A. Simon, *Models of Man* (New York: Wiley, 1957).

7. M.H. Segall, D.T. Campbell and M.J. Herskovits, *The Influence of Culture on Visual Perception* (Indianapolis: Bobbs-Merrill, 1966), 24.

8. See for example, K.E. Boulding, *The Image* (Ann Arbor: University of Michigan Press, 1956); and D. Lowenthal, "Geography, Experience, and Imagination: Towards a Geographical Epistemology," *Annals of the Association of American Geographers*, LI (1961), 241-260.

9. W. Kirk, "Historical Geography and the Concept of the Behavioral Environment," *Indian Geographical Journal*, Silver Jubilee Edition, (1951).

10. J. Wolpert, "Behavioral Aspects of the Decision to Migrate," *Papers and Proceedings, Regional Science Association*, XV (1965), 159-169.

11. See D.W. Harvey (1967), *op. cit.*

66

12. P.B. Warr and C. Knapper, *The Perception of People and Events* (New York: Wiley, 1968), 2.

13. Warr and Knapper (1968), *op. cit.*

14. M. Fishbein, "Attitudes and Prediction of Behavior," in M. Fishbein (ed.), *Readings in Attitude Theory and Measurement* (New York: Wiley, 1967), 483.

15. C.K. Ogden and I.A. Richards, *The Meaning of Meaning* (New York: Harcourt, Brace and World, 1930), 78.

16. C. Morris, *Signification and Significance* (Cambridge: M.I.T. Press, 1964).

17. In psycho-linguistics I am referring to C.E. Osgood, C.J. Suci, and P.H. Tannenbaum, *The Measurement of Meaning* (Urbana: University of Illinois Press, 1957); and R. Rommetveit, *Words, Meanings, and Messages* (New York: Academic Press, 1968); the perception material is summarized in Warr and Knapper (1968), *op. cit.*

18. See Warr and Knapper (1968), *op. cit.*, 7-13.

19. J.S. Bruner, J.J. Goodnow, and G.A. Austin, *A Study of Thinking* (New York: Wiley, 1956), 9.

20. See B.L. Whorf, *Language, Thought, and Reality* (Cambridge: M.I.T. Press, 1956); and Segall, *et al.* (1966), *op. cit.*

21. W.S. Torgerson, *Theory and Methods of Scaling* (New York: Wiley, 1958).

22. D.E. Berlyne, *Structure and Direction in Thinking* (New York: Wiley, 1965), 50-51.

23. P. Gould, "On Mental Maps," *Michigan Inter-University Community of Mathematical Geographers*, Discussion Paper No.9, (1966).

24. I think the kind of research design developed in Segall, *et al.* (1966), *op. cit.*, provides an interesting model, while J. Sonnenfeld, "Environmental Perception and Adaptation Level in the Arctic," in D. Lowenthal, (ed.) "Environmental Perception and Behavior," *Research Paper 109, Department of Geography, University of Chicago* (1967), provides a geographical example.

67

25. Osgood, *et. al.* (1957), *op. cit.;* Warr and Knapper (1968), *op. cit.;* and R.M. Downs, "Approaches to, and Problems in, the Measurement of Geographic Space Perception," *Bristol Seminar Papers,* Series A, No. 9, (1967).

26 K. Lynch, *The Image of the City* (Cambridge: M.I.T. Press, 1960).

27. A. Kaplan, *The Conduct of Inquiry* (San Francisco: Chandler, 1964) 62-71.

28. W. Isard and M.F. Dacey, "On the Projection of Individual Behavior in Regional Analysis," *Journal of Regional Science,* IV (1962), 1-32 and 51-83.

29. For example, R.W. Kates, "Hazard and Choice Perception in Flood Plain Management," *Research Paper No. 78, Department of Geography, University of Chicago* (1962); T.F. Saarinen, "Perception of Drought Hazard on the Great Plains," *Research Paper No. 106, Department of Geography, University of Chicago* (1966); Gould (1966), *op. cit.,* and P.R.Gould and R. White, "The Mental Maps of British School Leavers," *Regional Studies,* II (1968), 161-182.

THE TRANSITION TO INTERDEPENDENCE IN LOCATIONAL DECISIONS

Julian Wolpert
and
Ralph Ginsberg*

University of Pennsylvania

A. Empirical Context

As the circumferential highways have largely completed the
relatively simple task of laying a carpet through areas of
suburbia and beyond, there has been increased pressure to complete the final phase of the federal highway program -- the
linking of such rims to center cities.   The timing had been
unfortunate for the highway planners, for in the interim period,
the likelihood of a noncontroversial selection of metropolitan
rights-of-way has evaporated.   The period coincided with the
evolution of potential opposition groups which quickly learned
about the acquisition of power.   Simultaneously, there has
been open skepticism about the value of such highways in contrast to mass transit systems, especially when the highway

_____

    * The authors are associated respectively with the
Regional Science Department and the Sociology Department, University of Pennsylvania.   Grateful acknowledgement is made to
the Social Science Research Council, Committee on Governmental
and Legal Processes for their research support.

proponents appear to constitute a conspiracy by cement, tire
and automobile manufacturers. The controversies have extended
to New York, Philadelphia, Boston, Baltimore, Nashville and
Washington, D.C., and indeed to every other major urban center.
Although these cases differ in detail, the common pattern is
clear enough to permit general discussion.

Interest in the controversies is, however, not confined
to the politicians and planners who must deal with them in
order to implement the final stages of complicated plans in
which substantial financial and political capital has been in-
vested. They are of interest also to location theorists
since they call into question the validity of the classical
economic model on which the plans are based. We have in mind
especially assumptions about the independence of benefits,
costs, and actions of the relevant groups and the static character of the classical models. In this paper we develop some
of these implications for location theory and indicate some of
the ways in which location theory must be modified in order to
treat them adequately. Although expressways are here emphasized, we feel that the problems are endemic, and that conflicts over the location of schools and other private and public development programs would serve our purposes just as well.

B. The Development of Highway Controversies

Public concern with highway programs in all major urban
centers, and the activities of groups which these programs
specifically affect, seem to go through a common developmental
sequence. At the first stage of planning urban expressways,
the routes are selected by the traffic engineers seeking a
minimum cost or maximum flow connection from the rim highway
to a center city point. Such networks aid suburbanites,
trucking firms and center city development relatively the most.
Routes may follow the most inexpensive acquisition right-of-
way and, therefore, slice through the black ghetto zone,

assisting renewal in such core areas as a welcome but secondary
result. The program is typically unilateral in its conception
and too expensive in its study phase to be considered merely a
suggested route. If public hearings are held, they are often
put together hurriedly and furtively with agendas determined by
the policy officials and with responsibility assigned to a
bureaucratic catacomb. This is good strategy for preserving
independence, prerogative and discretion. Opposition groups
learn about the hearings too late, cannot penetrate the
officialdom to locate those responsible for the plan and do
not have the resources for drafting an alternate solution.
Neither are they experienced in the techniques of political
influence. In this stage benefits and costs are largely
imposed on affected groups. Furthermore the costs imposed on
some affected groups, especially those living in areas ripe
for "urban renewal," are often greater than the benefits they
receive through compensation and services, thus motivating
them to try to block the plan if they can.

The second stage, if it occurs, is typically a stalemate
-- the power imbalance begins to equalize or the remoteness
and obscurity of the policy-making strategists is partially
overcome, and the issue is defined. Publicity in the press,
alliances with well-meaning citizens, the involvement with the
courts, the threat of physical boycott or violence are all attributes of the stalemate period. This too may be the last
stage, and the artifact of the conflict might be only the
faded blueprint of the original plan and the skeleton of an
opposition group ready to be revived when a new threat emerges.
The policy-makers, at worst, suffer a modest setback and
emerge relatively no worse off than the opposition groups.
Such a stalemate is, however, also an early stage of *recognition of the interdependence* of the affected groups and policy-
makers involved. The stalemate is instructive in other ways

for it may lead to recognition that by cooperating and providing inducements as well as punishments, all groups can be
better off.

The stalemate may be broken in a third stage if an equali-
zation of power permits the abandonment of the original highway
plan in favor of an alternative acceptable to opposition groups,
i.e. restudying alternate routes or designs. The result of
this stage might be a compromise route or plan, as well as a
more sophisticated and interdependent decision-making procedure which may carry over into other issues and affect their
course. This stage reflects the confidence of opposition
groups in departing from the stalemate stage, fully able to
retaliate when confronted with the possibility of imposed
costs. This period of naive distrust is a prerequisite for
the lock-in on cooperative behavior, which is the new equilibrium point of "enlightened trust." There can be stability
in this stage as well but the precedent of interdependency has
been established. Added to the preliminary planning of the
engineering and economic aspects of highway construction are
then the predictions of conflict and evolution, for it may be
more efficient in the long run to design initially a stage
three compromise system to ensure its success.[1]

C. Extensions of Locational Analysis

There then appears to be a whole class of locational issues for which our traditional locational theory tools are
inapplicable beyond a very general level. A class of issues
are involved rather than a collection of unrelated exceptions,
for it is the presence of parallel attributes in a great number of situations that encourages us to suggest a more general
theoretical framework. This class of situations all involve
interdependent activities of the various groups involved.
Situations, like stage one of the highway conflict, in which
locational decisions are unilateral or independent, may then

be viewed as a special case of the more general approach of interdependent structures.

In the early stages of location research, the emphasis has been on normative models of rational decision-making based on the twin assumptions of utility maximization and the independence of the participants in the decisional process. Subsequent work, using a behavioral approach, has modified this model by taking account of the multiple goals and restricted capacity of decision-makers which limit the applicability of maximization or least cost solutions as these are usually understood. We now suggest that more serious limitations of the normative model stem from the fact that the participants in locational decisions are often groups whose relative organization, sophistication and power *develop* through the decision process, and whose interests are often conflicting but *interdependent*. In such game-like situations, which can be broadly conceived as bargaining structures, the rationality assumptions of classical economic theory break down.

These locational issues have been becoming increasingly prominent because of the greater visibility of the whole decision-making process and the greater participation of individuals and groups in the location of facilities or other institutions. This has had the effect of undermining the traditional market mechanism or narrow efficiency criteria according to which independent solutions could be safely assumed. Thus, a great many issues previously considered to be the prerogative of policy-makers legitimately entrusted with discretion for making definitive and independent decisions, have been transformed by the inclusion of other groups with legitimate claims into interdependent events which involve controversy or conflict.[2] Conflicts of interest are then the necessary conditions for the transition from independent to interdependent status and the sufficient conditions are

provided by the actual and anticipated distribution of *power*,
if power may be defined as the ability to alter a decision or
achieve a favorable outcome.   While there are strong incentives for policy-makers to encourage this transition in participation provided that wider participation increases the likelihood of cooperative solutions and the greater rewards to be
obtained thereby for all concerned, if threats are exercised,
as in the stalemate stage, the result will be detrimental to
everyone, including the policy-makers.   Two questions then
immediately arise: what induces policy-makers to admit the
legitimacy of the claims of other groups (thereby accepting
interdependent decision-making)? and what are the forces pro-
ducing cooperation rather than the mutual, destructive, exercise of threats?

D. Illustrative Gaming Structures

The most general and powerful instrument which has been
developed to analyze interdependent decision-making structures
is the n-person game.   Games may be set up to model situations
ranging from pure conflict to pure cooperation, and including
so-called mixed-motive or bargaining structures in which the
players are motivated to cooperate to achieve mutually beneficial outcomes but to compete over how the benefits of cooperation are to be distributed.   In bargaining games, which
correspond to the "imperfect" (oligopolistic) markets of
economies, any one player's actions are assumed to affect the
outcomes of the other players and, furthermore, rational players who try to gain as much for themselves as they can, are
assumed to be aware of this fact.   The highway location issues
described above are of the mixed-motive bargaining type.[3]

One simplified interpretation of the events now taking
place in conflicts over locational decisions may be illustrated
by means of the well-known Prisoners' Dilemma pay off matrix
(Figure 1).[4]   For purposes of simplicity, we consider only two

"players": the *policy-makers* (those legitimately entrusted with making locational decisions) and the *impacted population* (those directly affected by such decisions), and two choices available to each, a "cooperative or conciliatory" strategy (C) and a "defecting or refractory" strategy (D). The generalized pay-off structure (Figure 1b) (to use Rapoports picture-esque mnemonics) includes the reward for bilateral cooperation (R), the temptation (T) and sucker's pay-off (S) and the punishment (P) for bi-lateral defection. (The first pay-off in each cell is for the policy makers, while the second goes to the impacted group.)

Figure 1

Impacted Group

		C_2	D_2	C_2	D_2
Policy-Makers	C_1	5,5	-3,3	R_1,R_2	S_1,T_2
	D_1	10,-10	-5,-5	T_1,S_2	P_1,P_2
		(a)		(b)	

Now consider our highway example. In the preliminary period, a unilateral stage of final discretion by policy-makers with pay-offs indicated in the D_1 - C_2 box (a pay-off of ten units to the policy-makers who "defect" (D) at the expense of the impacted group who "cooperate" (C). This would suggest a period of D_1 - C_2 plays or "sucker" pay-offs for the impacted group measured in terms of the inconvenience or harmful effects of locating an undesired facility in the proximity of that group or displacing them. The positive pay-off to the policy-makers is perhaps a measure of the heightened power or status resulting from its ability to carry forth a unilateral decision. Thus, while the impacted group actually may choose between the alternatives cooperation or defection, they lack the insight and organization to take advantage of the opportunities built

into the matrix (their power).   By default, they choose the
conciliatory strategy and assume the sucker's role.   The pay-
offs are made asymmetrical to reflect the greater vulnerability
of the impacted group.

A second phase results after a marshalling of forces by
the impacted group in response to the desire to discontinue the
sucker's position.   At this point (stage two) it exercises
some of its retaliatory power.   This stage is represented by
the greater symmetry of pay-offs to the participants especially
in terms of the increase of T_2 so that it is now greater (i.e.
more positive) than R_2.   (Figure 2).   The impacted group can
now retaliate by selecting the defecting choice with consequences less severe than the martyrdom position.   Thus, the
D-D outcome may occur frequently within this stage until --
and if -- the mutually harmful effects lead to the third stage
of bilateral cooperation (C-C).   If only one participant decides to cooperate, the likelihood is that this C-D or D-C
move will lead on the following play to D-D because the defecting player will wish to continue selecting that choice which
resulted in the high pay-offs and the martyred player will want
to retaliate.   Thus, only the simultaneous decision by both
to cooperate can remove the participants from the D-D trap.
There is some stability in this more balanced relationship.

Figure 2

	C_2	D_2
C_1	5,5	-10,10
D_1	10,-10	-5,-5

It is interesting to note that in each of the three stages
(D-C, D-D, C-C) the players' expectations are mutually respon-
sive and the strategy based on these expectations is reasonable.
Only in stage three, however, are they reasonable for the

right reasons. In stage one (D-C) the impacted group plays
the role of sucker and the policy-makers take advantage of
them largely because the impacted group does not appreciate
its potential power and neither group is fully aware of the
(nonmonetary) costs for the impacted group generated by the
plan. In stage two (D-D), as in Cournot's price variation
oligopoly, both players receive low payoffs because, assuming
the other is behaving in a short-sighted noncooperative way,
they plan for the worst and thereby make it happen. Only in
stage three (C-C) do they aim for a cooperative solution and
achieve their real objectives.

At each stage, because the strategies of the players' are
adapted to one another, there is a relatively stable basis for
predicting results. Only in stage three, however, are expectations and behavior mutually and validly adjusting. Stages
one and two, as it were, are unstable in the long run and contain the seeds of their own undoing. The problem becomes one
of accounting for the timing and manner of transition between
stages. If we define a structure as *legitimate* when expectations are mutual and reciprocal, each stage of the highway
controversy has its own legitimacy. There is much evidence to
suggest that the stability of expectations which permitted, for
example, the highway engineers to locate their new expressway
according to purely economic criteria has been undermined.
Critical locational decisions are now being forced into the
courts and a new basis for legitimacy in making such decisions
is being forged through changes in power alignments. That is,
locational issues heretofore considered to be of almost unilateral discretion are now prime sources of conflict and the
outcomes of such conflicts are not easily predictable because
of institutional changes. When the black business district
and community of Nashville and Washington were threatened with
disruption because of urban expressways designed to ease the

flow of suburbanites to the central cities, the legitimate
policy-makers were surprised by the ensuing court struggle.

The transition from one stage to another clearly requires
relearning of expectation and orientation and does not occur
all at once. Rather, many halting but unsuccessful attempts
at cooperation (trial-and-error type learning) probably precede
any stable, mutually beneficial exchange. Especially when
there is strong situational stress resulting from high threat
or exogenous factors (e.g. ideological changes in society as a
whole), initial cooperation is likely to be unstable. Such a
"crisis" situation is exemplified by the classic theater fire
panic.

If we consider the problem, for example, of a population
seated within a theater having some distribution of exits, we
may think of these exits as a finite number of alternatives.
The selection of an exit or alternative by an individual is
typically considered not very salient or as a matter of indifference under normal circumstances. However, if a fire
were to suddenly start in one section of the theater, then the
selection from among available alternatives becomes quite important. In addition to the uncertainty to be derived from
the physical environment, i.e. the rate of the spread of the
fire to the individual and to the exits, there is added the uncertainty of interdependency with the decisions of others
which will affect his rate of speed to the various exits.
The *a priori* institutional setting might lead the decision-
maker to assume a rational queuing process to the exits with a
priority established by relative position. The presence of
the fire has the effect, however, of introducing ambiguity
about the maintenance of an orderly queue based upon position
priorities, for those whose queue position mitigates against
survival may attempt to overtake the others and become surviv-
ing cowards rather than burned heroes.

Disturbances to the priority queue upsets the institutional framework and undermines the basis of expectations in
this interaction process. Panic fed by the likelihood of
being trapped and by contagion and uncertainty about the behavior of others, has the effect of "tunnel vision," the reduction of alternatives to a single prominent choice which is
most conspicuous. In this case, the ambiguity is clearly
dysfunctional and the increased errors in the choice behavior
of those in panic is harmful relative to those not in panic
who remain adaptive. This strategy may indeed be quite effective and constitute an important tool in the kit of experienced
negotiators and politicians -- that is to carefully control the
outflow of communications to one's opponent in order to effect
ambiguous signals, unless, of course, this dangerous strategy
has effects which are fed back to its originator.

E. Conclusion

In the foregoing discussion we have described a series of
stages through which we think highway controversies, and indeed
many other controversies over location of public facilities,
pass. We have indicated why, except at the very beginning of
the process, traditional location theory is inadequate as a
descriptive and normative tool. The participants in the
locational decision are seen to become increasingly interdependent. As a step toward greater realism, highly structured
but still simple models which specifically incorporate interdependent actors are sketched. Clearly these models are only
a beginning: they define the dimensions of the problem. A
greater number of alternatives and participants[5] need to be
introduced, the bargaining process must be looked at from a
more microscopic and dynamic perspective, and the forces leading to transition from one stage to another must be more
clearly specified. Perhaps then, in empirical analysis we

can get beyond the apposition of apparently similar cases. In research now underway, which will be reported upon elsewhere, the authors are investigating these problems.

NOTES

1. Perhaps the notion of "the public interest" requires that cooperative solutions be *imposed*, but this may entail a considerable added "public cost."

2. Distinction must be drawn between (1) interdependency in connection with one issue because of *bargaining* by the participants and (2) interdependency because many issues are treated as a package rather than separately (independently). Focus of attention is reserved here for the type (1) process but in further research now underway an analysis is being made of facility complexes and their complementary trade-offs.

3. General characteristics of this category of games are well described in R. Duncan Luce and Howard Raiffa, *Games and Decisions,* (New York: Wiley, 1957).

4. Refer to Anatol Rapoport and Albert Chammah, *Prisoner's Dilemma,* (Ann Arbor: University of Michigan Press, 1965) and to J. Wolpert, *et al.,* "Learning to Cooperate," Papers, *Peace Research Society*, VII, (1966).

5. See J. Wolpert, *et al.,* "Coalition Structures in Three-Person Non-Zero-Sum Games," Papers, *Peace Research Society,* VII, (1967), 97-108.

A "FRIENDS-AND-NEIGHBORS" VOTING MODEL AS A SPATIAL

INTERACTIONAL MODEL FOR ELECTORAL GEOGRAPHY

David R. Reynolds

University of Iowa

The purpose of this paper is to examine one approach to a
fundamental problem confronting the geographer interested in
developing models of voting behavior -- that of more satisfactorily integrating spatial and behavioral approaches given the
types of data availability constraints typically encountered
in electoral geography.   In an attempt to overcome this problem, a simple model is developed and tested empirically.
Results suggest that derivatives of the model have considerable potential in analyses of political processes over space,
particularly those directed toward discerning the impact of
localisms on a voting response surface.

A. Introduction

Although voting data collected by areal units are among
the types of quantitative data most available to the researcher
they have not seemingly afforded the basis for studies in
which interpersonal interaction over space is a significant
factor.[1]   Presumably this is one reason why the vast majority
of studies of voting behavior and political participation have
not adopted a more spatial approach.   Contrariwise, in such

fields as locational economics and economic geography, a more
spatial approach in research has been facilitated by the
availability of data descriptive of economic activity for
points which can be designated by relatively unambiguous locations and because of the availability of data descriptive of
flows between such points. Despite the absence of flow data
relevant for most political activities, it should be apparent
that spatial effects and interactions in large part generate
any voting response surface. This becomes more obvious when
one considers that the individual's voting decision is to some
extent dependent upon his access to information regarding the
candidate or issue, which in turn is partially dependent upon
his relative location in a communication network within social
groups.

To elaborate on the above point further, there is evidence which strongly suggests that direct personal contacts of
a lasting and intensive nature are primarily established between individuals separated by short physical distances.[2]
Furthermore, the groups in which a large number of the individual's social relations take place, such as acquaintance circles, work groups, and other "primary groups" would also appear, to a great extent, to be based on short distance interactions. Once an individual has become a group member, his
membership will further reinforce his propensity to interact
over short distances if he is to remain an active participant
in the spatially concentrated group. Thus, it would appear
that a very large part of the individual's experiences is
derived from, and has reference to, his immediate surroundings
from which he will also take over many of his political attitudes, beliefs, norms, habits, and ways of behavior.

Despite the empirical basis and logical appeal of the
above comments, the literature on voting behavior, geographical
or otherwise, is devoid of models for evaluating the impact

of space upon political process adaptable for empirical testing
with the aggregate data so readily available.   This is not
meant to suggest that the problem is merely one of data availability.   Before the researcher hastens to collect more behaviorally and spatially pertinent information on individuals, he
should perhaps be more certain that he has not failed to construct models the empirical testing of which extracts the
maximum possible behavioral information from aggregate data.
The model outlined here was developed with this purpose in
mind.

B. A Neighborhood Effect in Voting Patterns

The model to be presented in this paper is based in part
upon the premise that there is a neighborhood effect in voting
patterns isomorphic to those illustrated by Hägerstrand and
others for the diffusion of innovation.[3]   In support of such
a premise the late V.O. Key, Jr. has shown that in most governmental elections candidates receive more votes from their home
districts or from groups with which they are identified (i.e.,
religious, ethnic, etc.) than from the electorate as a whole.
He refers to the first of these incremental differences in
votes as "friends and neighbors" voting and has shown via maps
(based on county voting statistics) how widespread it is in
local and state elections in the southern states of the United
States.[4]   The candidate's major strength comes from his home
county or precinct and there apparently is a strikingly regular decay in support with increasing distance out from the
home location.   In rural and small town areas, Key found this
tendency to be particularly strong-- in these, voters are more
likely to know the candidate personally or to know someone who
does.   Such a phenomenon is also observable in urban areas
although the decay with increasing distance is more pronounced,
presumably the result of increased population densities and

84

the higher probability of information travelling over shorter
distances.

In two studies directed toward an explanation of areal
variations in voting, this neighborhood or "friends-and-neighbors" effect is also apparent.   In the first instance reference is made to a study by McCarty of the distribution of the
relative number of people per county who voted for the late
Sen. Joseph M. McCarthy in the 1952 Congressional election.[5]
By applying simple regression analysis, McCarty tested the hypothesis that the percentage of the total population per county
which was rural was areally associated with the McCarthy vote.
Upon examining the spatial pattern of residuals from this regression, McCarty observed an apparent distance decay in the
residuals in all directions from the home town of the Senator.
After the inclusion of a second independent variable, distance
from the center of a county to the Senator's home town, in the
regression analysis, the amount of variation in the McCarthy
vote "explained" by regression increased by thirty-two per cent.

The second of these two studies is that conducted by
Roberts and Rumage.[6]   Somewhat more ambiguously than in the
aforementioned, a neighborhood effect in the distribution of
left-wing voting in England and Wales was isolated.   Roberts
and Rumage hypothesized and found a distance decay in left-
wing voting away from coal fields (which are presumably "poles"
of left-wing voting strength).   In this instance, the neighborhood effect is unlikely to be of the "friends-and-neighbors"
variety as defined by Key.   Nonetheless, it appears to be a
larger scale analogue based upon the flow of political information from areas of nucleated class and party consciousness.

These studies all suggest that the impact of a neighborhood effect in voting (perhaps better termed in-group voting)
decreases as the areal extent and membership of an in-group
increases.   Thus far, no evidence has been found to refute

this. In fact, maps depicting the areal distribution of support for competing parties or candidates often reveal apparent
evidence of such spatial regularities which sometimes manifest
themselves in marked regional differences in voting behavior.[7]

However, it cannot always be expected that a neighborhood
effect will be readily apparent from an areal distribution of
election returns. Although probably operative in all democratic elections, this type of spatial regularity would be most
pronounced in intra-party political primaries for state level
offices or in contests where the majority of voters do not have
strong party loyalties and where one or more of the candidates
are seeking major political office for the first time.   In
successive campaigns such candidates will attempt to "flatten"
the "distance decay" in their voting support (without losing
the support of their "friends-and-neighbors") by generating
issues and aligning with nonlocally based organizations which
cut across the population as a whole.   In affect a candidate's
localized support probably does not diminish over time: instead, it becomes blurred and sometimes over-shadowed by his
development of cross-cleavages over an area.   The importance
of a candidate's developing such cleavages has been documented
by several authors.[8, 9]

In the following section a simple formalization of in-
group voting in a spatial context is presented.   It is essentially a correction, modification, and spatial extension of a
model presented and tested by Coleman.[10]   Unlike Coleman's
model, however, it provides an example of modeling political
behavior utilizing voting data aggregated by areal units.

C. Model: Voting for an In-group Member as a Function
of Social and Spatial "Friction"

Consider a two behavior state situation in which some
members of a bounded areal group (e.g., residents of a precinct, county, or groups of contiguous voting districts, etc.)

vote for a candidate without regard to areal group affiliation
(State I), while other members always vote for a candidate who
is a member of the areal group solely on the basis of his membership in the group. It is important to note that members
in State I may still vote for a candidate from the areal group
although they do so on some basis other than his territorial
identification or areal group affiliation. Examples of State
II behavior would be (1) when the voting decision is predicated upon close interpersonal interaction between the voter
and the candidate or friends of the candidate or (2) when the
voting decision is predicated upon a real or imagined bond
between the individual voter and the candidate, exemplified by
responses wuch as "I voted for candidate X because he under-
stands the problems of *our* area."

When the voting decision is based primarily upon any type
of consideration other than interpersonal interaction or territorial loyalties, it is considered to be State I behavior.
As can be seen, State I is the more all-inclusive voting behavior category.

Assume that there is some transition rate q_{12} from State
I to State II which is essentially composed of "random shocks"
resulting from variables not included in the model. Also
assume that there is a transition rate from State II to State
I, say q_{21}, which is composed not only of "random shocks" but
also of a linear function of S_1 and S_2 (see the diagram of the
process below)

where:

> S_1 = impediment to interaction which reflects
> the increasing difficulty of communicating
> over greater distances as necessitated by
> increases in the *areal* size of the group

and

> S_2 = impediment to interaction which is due to
> the increasing complexity of the group as
> its membership increases.

State I	State II
(Vote independently of membership)	(Always vote for member)
n_1 = number of individuals in this state	n_2 = number of individuals in this state

$$\xrightarrow{\hspace{2cm}} q_{12} = \varepsilon_1$$

$$q_{21} = \alpha S_1 + \beta S_2 + \varepsilon_2 \xleftarrow{\hspace{2cm}}$$

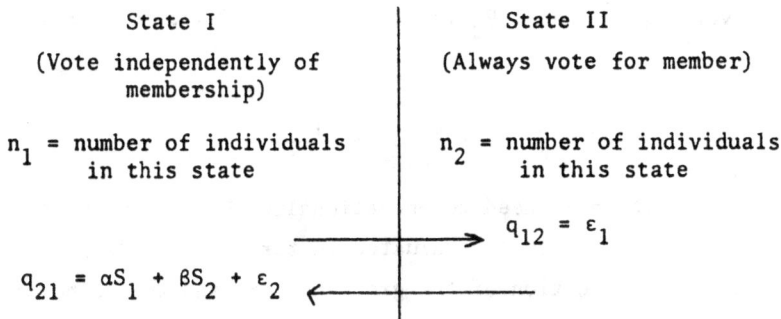

There are n_1 and n_2 individuals in the two states respectively. The numbers of individuals in each state need not be constant proportions of the total group membership (i.e., total voting population of areal units or units under consideration). Indeed, from the above assumptions, it may be seen that the rate of change in the size of n_2 with respect to time may be expressed as

$$\frac{\partial n_2}{\partial t} = \varepsilon_1 n_1 - (\alpha S_1 + \beta S_2 + \varepsilon_2) n_2$$

At statistical equilibrium, which seems to be a reasonable assumption in most voting situations,[11] the number of individuals shifting between States I and II will be equivalent, so that:

$$\varepsilon_1 n_1 = (\alpha S_1 + \beta S_2 + \varepsilon_2) n_2$$

If the term $\varepsilon_1 n_2$ is added to *both* sides of the equation then

$$\varepsilon_1 (n_1 + n_2) = (\alpha S_1 + \beta S_2 + \varepsilon_1 + \varepsilon_2) n_2$$

or

$$\frac{n_2}{n_1 + n_2} = \frac{\varepsilon_1}{\alpha S_1 + \beta S_2 + \varepsilon_1 + \varepsilon_2}$$

which is seen to be the proportion of the total group voting for the candidate because he is a group member. Let this proportion equal P_2. Then

$$P_2 = \frac{\varepsilon_1}{\alpha S_1 + \beta S_2 + \varepsilon_1 + \varepsilon_2}$$

or, inverting so that $1/P_2$ is an increasing function of S_1 and S_2

$$\frac{1}{P_2} = \frac{\alpha}{\epsilon_1} S_1 + \frac{\beta}{\epsilon_1} S_2 + \frac{\epsilon_1 + \epsilon_2}{\epsilon_1}$$

Clearly there is need to operationalize P_2 because neither n_1 nor n_2 can be directly evaluated in terms of the data. However, the proportion of the group members and the proportion of the *total electorate* voting for the candidate are easily available (depending, of course, upon the areal units of observation). Labeling these proportions as p and t, respectively, P_2 can be approximated as p - t. This seems to be a reasonable approximation, since P_2 is not the proportion of the areal group members voting for the fellow group member, but rather the proportion voting for him in excess of those who would be expected to vote for him anyway. Therefore,

$\frac{1}{P_2} = \frac{1}{p - t}$. Now, in order to ascertain the empirical nature of the relationship, it becomes necessary to operationalize the variables S_1 and S_2. It seems reasonable to define

$$S_1 = D; \qquad S_2 = \ln N$$

where D = distance (time, airline, road, etc.) from the center of the areally defined group (candidate's headquarters, for example) to the boundary of the district, and N = the size of the group (voting population). Each of these would seem to be an appropriate *first* approximation to the friction of interactions within the areal groups. The assumption of a log linear relationship between q_{21} and N says that the increment in within-group voting on a friends-and-neighbors basis added by a new member when the group is of size 100 is only one-tenth the increment added by a new member when the group is of size ten.[12]

It is now possible to empirically examine the effects of interactional frictions upon the propensity of areally defined groups to vote for a group member candidate through the use of least squares estimating procedures.

$$\frac{1}{p - t} = aD + b \ln N + c$$

where $1/(p - t)$ is defined as above and

$$a = \frac{\alpha}{\varepsilon_1} \qquad\qquad b = \frac{\beta}{\varepsilon_1} \qquad\qquad c = \frac{\varepsilon_1 + \varepsilon_2}{\varepsilon_1}$$

Since parameters a, b, and c are ratios, it is not possible to arrive at independent estimates of α, β, and ε_1. Nevertheless, a, b, and c have significant behavioral interpretations of interest to the student of group political behavior. The parameter a would be interpreted as the rate of increase with increasing areal size of the group (e.g., voting district) in the mean tendency to vote independently of group membership, relative to the tendency to vote on the basis of group (locality) membership (group size held statistically constant); b would be interpreted as the rate of increase per added areal group member in the mean tendency to vote independently of group membership, relative to the tendency to always vote for a group member (areal size of the group (district) held statistically constant); and $1/c$ would be interpreted as the initial (group size of two and zero distance) average tendency to always vote for a group member.

Empirical examination of this formulation should contribute to our understanding of neighborhood effects upon the electoral process; and, in particular, contribute towards a more incisive examination of the effects of areal and population size upon the development of "localism" or "regionalism" in voting patterns. Time series analyses of this formulation should also contribute to our knowledge of political process over space -- particularly the oft-noted breakdown of localisms over time.[13]

D. An Empirical Example

In a current research project, the author is utilizing a derivative of this model in an examination of the results of spatial competition between candidates in the 1954 Democratic gubernatorial primary in Georgia. This particular contest was selected primarily because of the availability of complete election returns and population data at the precinct level.[14] But it was also an intra-party contest in which three of the five major candidates were running for state level offices for the first time -- a situation in which "friends-and-neighbors" voting is most likely to be evident. Initial findings have been most encouraging; some of them will be presented by way of example below. However, the final results of this research will be reported at a later date.

The two leading candidates together polled slightly over sixty-one per cent of the popular vote in the primary. The winner was the incumbent lieutenant governor while the second candidate (in terms of voting support) was a former governor. Although one component element of the voting response surface for each of these candidates as a friends-and-neighbors effect, this effect does not take the form of a regular distance decay out from their bailiwicks since each had affiliated himself with nonlocality based issues, partisan causes and quasi political organizations. However, the third leading candidate, Thomas Linder, although having held an appointed office for several years, was running for a major political office for the first time. It was, therefore, expected that his voting response surface would be dominated by a neighborhood or friends-and-neighbors effect. Since this candidate's polling strength was limited to his home county and surrounding

counties, a reasonably compact, arbitrarily bounded, area composed of 248 contiguous precincts centered on the candidate's
"home town" precinct was selected as a sub-study area.
Although the candidate received only 13.50 per cent of the
state-wide popular vote, approximately one-third of his total
voting support was concentrated in this predominantly rural
area.

In order to roughly assess the degree to which the voting
proportions for Linder at the precinct level exhibited spatial
contagion, Dacey's generalization of the standard normal contiguity ratio for intervally scaled data was computed.[15]
After performing a logarithmic transformation on the voting
proportions (so as to meet the assumption of normally distributed data demanded in the calculation of the ratio) a contiguity ratio of 13.58 was obtained. Since this ratio is a
standard normal variate, the areal distribution of Linder's
voting proportions was judged to be significantly different
from random at a very high level. This finding gives further
credence to the suggestion that a spatial variable or a variable that varies systematically over space may account for the
observed distribution of voting proportions, although it, of
course, gives no indication as to what form the spatial regularity takes. It has already been suggested that the conditions of this contest might lead one to expect a neighborhood
effect of the form specified by the model developed in the
previous section. Therefore, the model was calibrated and its
goodness of fit in this context assessed.

Initially the observations on the variables $1/(p - t)$, D,
and N were to be provided by continuously incrementing (aggre-
gating) contiguous precincts in a concentric manner. However,
preliminary analyses suggested that the tendency to vote for
Linder did indeed decrease in a linear manner as the length of
the principal axis of the areal group increased but with

definite directional biases (depending upon the directional orientation of the "group"). In these preliminary analyses contiguous precincts were uni-directionally cumulated out from the general vicinity of Linder's home precinct along major highway routes; with the addition of each precinct, the value of p - t, N and D (road distance from the center of Linder's home precinct to the midpoint of the precinct last incremented to the group) was calculated from the original precinct data.

As a result of the apparent directional biases referred to above, a polar coordinate system comprised of rings spaced at the equivalent of three mile intervals and comprised of eight equal angle sectors was superimposed on the sub-study area. The sectors were oriented so that at least one U.S. or improved state highway traversed the entire length of each sector. This was possible in the sub-study area, since the highway network has a radial pattern. The sub-study area and sectors are delimited in Figure 1. In a manner similar to that of the preliminary analyses, precincts were contiguously aggregated by ring for each of the eight sectors. A precinct was included in a particular ring and sector, if its midpoint fell within that ring and sector. Then, the variables p - t, D and N were measured sector by sector for "groups" of increasingly larger size in terms of both population and airline distance from Linder's home precinct to the periphery of the group. In effect, these variables are derived from the cumulative (or marginal) distributions for each sector, as is necessary in order to test the model in the given context.

As mentioned above, this illustrative example is part of a larger study directed towards an analysis of the results of political candidates competing for votes over space in a manner somewhat analogous to spatial competition between firms.[16] Therefore, when examining the decay in the friends-and-neighbors support of a candidate with increasing distance, one must

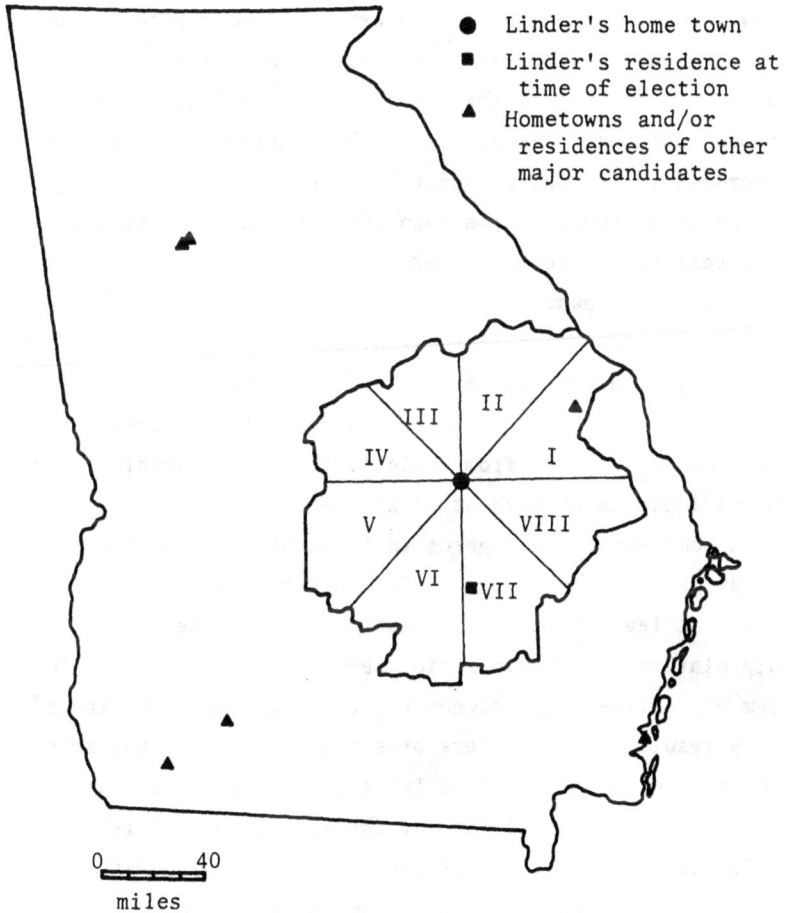

● Linder's home town
■ Linder's residence at
  time of election
▲ Hometowns and/or
  residences of other
  major candidates

Figure 1: The Sub-Study Area

consider where the focus of his support is relative to that of each of his competitors. The locations of the home towns and/ or long term residences of all the major candidates are indicated on the map in Figure 1. On the basis of neighborhood support alone, it can be seen that Linder faced more competition in the northern four sectors, particularly sector 1. His friends-and-neighbors support would appear to be stronger in the southern four sectors, particularly since his residence was in sector VII at the time of the election. (In fact, the preliminary analysis indicated that there was no discernible decrease in friends-and-neighbors support along the highway connecting Linder's home town with that of his residence). The testing of the model was therefore conducted in three stages as follows:

(a) for all sectors simultaneously (128 observations)

(b) for all sectors except I and VII (99 observations)

(c) for southern sectors except VII (54 observations)

As can be seen from Table I, the "goodness of fit" of the model improves at each stage as the sectors of analysis become more homogeneous as regards their locations relative to the locations of competitors. Column two of Table I indicates that, at least in this application of the model in a rural area of relatively even population density, there is a serious problem of collinearity between the two independent variables. As a result the parameters of a reduced form of the model are the only ones indicated in Table I. Nonetheless, as can be seen, the goodness of fit (of the reduced model) is still high (.730 for stage c). This indicates that in the study area there is a marked linear decrease in friends-and-neighbors support (or increase in the tendency to vote independently of areal group membership) as the distance from Linder's home precinct increases.

TABLE 1

Least Squares Evaluation of the Model

Stage	Coeff. of Determination	Correl. Between distance & log population	Coeff. of Determination	$\alpha \dfrac{\varepsilon_1}{\varepsilon_1}$	$\dfrac{\varepsilon_1 + \varepsilon_2}{\varepsilon_1}$
			for the model $\dfrac{1}{p-t} = \dfrac{\alpha}{\varepsilon_1} D + \dfrac{\varepsilon_1 + \varepsilon_2}{\varepsilon_1}$		
a	.596	.890	.540	.529	50.71
b	.789	.927	.699	.456	53.01
c	.938	.948	.730	.389	59.93

The collinearity problem present in this application of
the model, which does not appear to be serious in applications
where population density is highly variable for small areas,
prompted a slight reformulation of the model.   In the reformulation, the effect of population was deflated by dividing each
distance by the total population of the respective areal group.
The model then takes the form of

$$p - t = \frac{\varepsilon_1}{\alpha} \frac{D}{N} + \frac{\varepsilon_1}{\varepsilon_1 + \varepsilon_2}$$

where D and N are operationalized as before.   The estimate of
the parameter $\frac{\varepsilon_1}{\alpha}$ now takes on a different interpretation.   It
would be interpreted as the rate of increase in the mean ten-
dency to vote on the basis of group membership with increasing
"per capita communicating distance" between an individual and
the candidate's home town, relative to the tendency to vote
independently of areal group membership.   The resulting least
squares estimating equation for stage (c) is

$$p - t = 67.308 - 13.555 \frac{D}{N}$$

with an R^2 of .861.

This indicates that there is a very rapid diminution in
the tendency to vote on a friends-and-neighbors basis relative
to the tendency to vote independently of areal group membership as the "per capita communicating distance" is increased.
The very high correlation between distance and population
(.948 in Table 1) together with the fairly realistic assumption that information concerning the candidate is diffused
from the candidate's home town permits the operationalization
of $\frac{D}{N}$ as "per capita communicating distance."   However, in
other contexts, e.g., differing definitions of areal groups,
population densities, etc., such a variable may have little
substantive meaning.

Although only the areal size and population size have been incorporated into this model, there are no conceptual and few operational problems to the inclusion of measures of group heterogeneity, such as social class composition, in the model. In fact, the inclusion of other variables might serve to transform geographical space into a more politically relevant "space". The selection of other variables is likely to be dependent upon the political and national context of the voting situation(s) under examination.

F. Summary

The above empirical application of a spatial "friends-and-neighbors" voting model has been discussed at some length so as to present the researcher with an alternative method for analyzing aggregate voting data in a spatial context and yet one which remains free from many previous restrictions such as those encountered in nonspatial ecological analyses. This type of model is superior to the usual procedure of examining distance decay in voting proportions (e.g., in the McCarthy study) primarily because the basic parameters are behaviorally interpretable but also because the observational units are treated as potentially meaningful spatial groups rather than an unrelated collection of points or areas with voting behavior attributes to be explained statistically. Political processes operate in and usually over space. Models of these processes are likely to be more powerful when this dimension is given more explicit consideration.

98

NOTES

1. Kevin R. Cox, "A Spatial Interactional Model for Political Geography," *The East Lakes Geographer,* Volume 4 (December 1968), 58.

2. See, for example, studies of marriage and migration distances such as Richard Morrill and Forrest R. Pitts, "Marriage, Migration, and the Mean Information Field,: *Annals of the Association of American Geographers,* Vol. 57 (June, 1967) 401-422. Or, for a more general review article of further evidence, see G. Boult and C.G. Janson, "Distance and Social Relations," *Acta Sociologica,* Vol. 2 (1956), 73-98.

3. See, for instance, Torsten Hägerstrand, "On Monte Carlo Simulation of Diffusion," in W.L. Garrison and D.F. Marble (eds.), *Quantitative Geography,* Northwestern University, Studies in Geography, Number 13 (Evanston: Northwestern University Press, 1967), 1-31.

4. V.O. Key, Jr., *Southern Politics* (New York: Alfred A. Knopf, 1949).

5. Harold H. McCarty, "McCarty on McCarthy: The Spatial Distribution of the McCarthy Vote, 1952" (unpublished manuscript, Iowa City: Department of Geography, State University of Iowa, no date).

6. Michael C. Roberts and Kennard W. Rumage, "Spatial Variations in Urban Left-Wing Voting in England and Wales in 1951," *Annals of the Association of American Geographers,* Vol. 55 (March, 1965), 161-178.

7. See, for example, Stein Rokkan, "Geography, Religion, and Social Class: Crosscutting Cleavages in Norwegian Politics," in S.M. Lipset and S. Rokkan (eds.), *Party Systems and Voter Alignments* (New York: Free Press, 1967), 367-444.

8. See, for example, Erik Allardt and Pertti Pesonen, "Cleavages in Finnish Politics," in S.M. Lipset and S. Rokkan (eds.) *Ibid.,* 325-366.

9. The process by which candidates attempt to broaden the basis of their support over time can, not inappropriately, be likened to spatial competition between retail establishments and manufacturing plants where one firm extends its market area to the detriment of another by lowering delivered price, by more effective advertising, etc.

10. James S. Coleman, *Introduction to Mathematical Sociology* (New York: The Free Press, 1964), 269-277.

11. This assumption would have to be examined for the election under empirical consideration before testing the model.

12. Formally we are assuming that the change in q_{21} with respect to distance is constant and that the change in q_{21} per added group member is proportional to the reciprocal of group size:

$$\partial q_{21}/\partial D = \alpha \quad \text{and} \quad \partial q_{21}/\partial N = \beta/N$$

Then by integration of these two partials we have

$$q_{21} = \alpha D + \varepsilon_2' \quad \text{and} \quad q_{21} = \beta \ln N + \varepsilon_2''$$

Therefore the complete estimating equation for q_{21} is

$$q_{21} = \alpha D + \beta \ln N + \varepsilon_2 \quad \text{where} \quad \varepsilon_2 = \varepsilon_2' + \varepsilon_2''$$

13. Inasmuch as it is extremely unlikely that one would find monotonically increasing tendencies to vote independently of locality with increases in the areal size of electoral districts, a model similar to the above may contribute towards providing a partial solution to the problem of devising electoral districts to either promote or discourage localisms, depending, of course, upon the governmental nature of the political office under consideration.

14. The precinct voting data were obtained from Joseph L. Bernd, *Grass Roots Politics in Georgia* (Atlanta: Emory University, 1960).

15. See Michael F. Dacey, "A Review on Measures of Contiguity for Two and K-Color Maps" in Brian J.L. Berry and Duane F. Marble (eds.), *Spatial Analysis* (Englewood Cliffs, N.J.: Prentice-Hall, 1968), 479-495. And David Andersen, *Three Computer Programs for Contiguity Measures,* Tech. Report, No. 5, Spatial Diffusion Study, (Department of Geography, Northwestern University, Evanston, Ill., December, 1965).

16. The notion of spatial competition is particularly
appropriate since in Georgia primary victors are determined or
the "county unit system" -- a system similar to the U. S.
electoral college.

THE GEOGRAPHICAL RELEVANCE OF SOME LEARNING THEORIES

Reginald G. Golledge
Ohio State University

The search for explanations of the spatial behavior of
individuals and groups inevitably leads to a discussion of the
processes which influence behavior. Recent emphasis in geography on interaction, diffusion, and decision-making models,
and a surge of interest in some spatial aspects of psychophysical theories of perception confirm this trend. Another
process which involves some useful spatial concepts, but
which so far has merited scant attention in geography, is the
learning process. It is the aim of this paper to examine the
role of this process in spatial behavior, to indicate some
useful spatial concepts from learning theory, to review a
selection of learning models that could conceivably be used in
a spatial framework and to suggest some problems which are
suitable for analysis by the models presented in the paper.

    A. Learning Theory and Spatial Behavior

For the purpose of this paper "spatial behavior" is defined as any sequence of consciously or subconsciously directed
life processes which result in changes of location through
time.[1] "Learning" is the process by which an activity origi-
nates, or is changed through responding to a situation --

provided that the changes cannot be attributed wholly to matura-
tion or to a temporary state of an organism.[2]

Geographers have concerned themselves with a variety of
spatial behaviors. For example there exists a large volume of
cross-sectional studies of consumer behavior,[3] journey-to-
work,[4] production behavior,[5] and so on. Such studies examine
movement patterns through space, generally over a very limited
time period.

Perhaps the most frequently occurring reference to spatial
behavior in geographical literature occurs through the use of
"spatially rational man" assumptions. Just as classical economic theory postulated the existence of economically rational
man, so geographers have incorporated into their theories postulates which assume that man may have perfect knowledge of
alternative trip types, a stable system of ordered space preferences for goods and services, and a constant preference for
the least effort solution in trip making.[6] It is eminently
possible to theorize about such a man, and he has appeared consistently in geographic literature.

Recent papers have stressed that spatial rationality is
but one of a range of behaviors that actually occur in space.
For example, Wolpert[7] has examined the notion of bounded
rationality or satisficing behavior; Gould[8] has examined
search behavior; Golledge and Brown[9] have shown that both
search and stereotyped behavior can be observed at different
time periods in the marketing act; Huff[10] has attempted to
construct the motivational basis of spatial behavior; Marble[11]
has investigated asymptotic behavior in journey-to-work patterns; and Pred[12] has tried to construct a behavior matrix
depicting the dynamics of locational decision-making.
Techniques used in each study vary from choice models, to
Markov chains, renewal processes, and topology.

Problem-Solving and Habitual Behaviors

Each of these cases cited above show that the type of behavior with which geographers have most frequently concerned
themselves is problem-solving behavior. They have examined
such things as the decision processes that result in locational
choice; the problem of locating urban functions; the problem
of choosing paths to work, to shop and to play, and so on.
Once the problem appears to be solved, consequent behavior is
neglected. It is pertinent at this state to briefly examine
problem-solving behavior and its relation to other spatial
behaviors.

Problem-solving behavior is noticeably different from
weakly motivated (or random) behavior and is also different
from habitual response-making. It is characterized by:

a) confrontation with a problem

b) deliberate thinking in a specified direction
    (search, or vicarious trial and error behavior)

c) choosing among alternative courses of action
    and making overt responses.

Problem-solving behavior is a highly selective process and
represents the attempts of mankind to adapt to changing conditions. Changes in behavior due to problem solving are often
substantial and abrupt. Despite the apparent importance of
problem-solving behavior, however, it has been claimed that it
is not the most common form of human behavior.[13] Routine behavior or habitual response behavior is claimed to be more
frequent. Routine behavior generally follows problem-solving
behavior and is of greater duration through time. However,
if environmental changes are frequent, or if several alternatives of approximately equal magnitude of reinforcement are
faced, behavioral oscillation[14] may occur. The effect of
this activity on a model's predicting ability is obvious.
Routine behavior on the other hand often follows a path of

minimum effort, it serves to reduce uncertainty in the decision
process, and reduces consideration of alternative courses of
action.   In other words it is the behavior most used to cope
with the contingencies of everyday living.

Learning and Behavior

Regardless of the type of spatial behavior that is observed
at any time or that evolves through time, it is apparent that
such behavior is a learned phenomenon.   Consequently, some
knowledge of the learning process is essential to understand
it.   For example, the degree of correspondence between theoretical and observed behavior can be interpreted as a function
of the extent of complete learning in a given system.   It is
feasible to assume that at any given time, and for any existing
(noncontrolled) population, we can expect part of a population to exhibit forms of habitual behavior which might be
explained simply by least effort or other specified characteristics.   However, we can also expect a portion of the population to be in earlier phases of learning about the system in
which they live.   Recent in-migrants will exhibit some type of
spatially irregular search behavior; those partly acquainted
with the system in which their behavior is observed may have
partly formalized behavior; those with considerable experience
of the system may exhibit forms of habitual behavior.   We
can imagine that all members of the system are striving toward
some asymptotic spatial response.   Theoretically, this may be
complete spatial rationality; or it may be a series of habitual
responses that represent a satisfactory coping strategy, but
which are not limited to selection of a single alternative.
Approach towards this asymptote varies with the accumulation
of knowledge about any spatial system and extent of experience
with it.

B. Spatially Relevant Learning Concepts

Many concepts from learning theory have direct spatial application. By examining briefly a selection of theories, it is possible to isolate some concepts that appear useful for geographers.

Action Space

Recently geographers have found considerable value in the concepts of "life space," "action space," or "ego space."[15] Extensive use of these concepts was made by Kurt Lewin, who regarded behaving organisms as geometrical points moving about in life space.[16] Individuals were subject to the pushes and pulls of personal and group expectations. In the course of their locomotions through space, individuals circumvented barriers of one sort or another, and their locations at any point in time were determined by the forces impinging on their life space at that time. By adding the characteristic of mobility to each individual in his life space, Lewin envisaged movement towards locations that were adient (or which had a positive valence), and moving away from vectors in the life space that were abient (or which had a negative valence).[17] Ideas similar to this appear in geography in migration theory (e.g., the push-pull hypothesis) and the concept is implicit in the idea of town attractiveness, and the centrifugal-centripetal forces literature in urban geography.

The presence of adient-abient vectors in life space gives rise to conflict situations. Examples of these are:

a) where two positive valences of equal force exist (leading to an approach-approach conflict similar to the concepts implied in the selection of market centers)

b) where two negative valences exist (leading to avoidance-avoidance situations)

c)   where there are simultaneously present positive
     and negative valences (leading to the attraction-
     friction combination used frequently in geography.)

Lewin also argued that only where life space (which is perceptual) and physical space (which is actual) coincide, can
Euclidean-space distances and directions be used to describe
either the locomotions (paths) or the responses (movements) of
people.   Otherwise he argued that some type of space-trans-
formation should be used.   It is interesting to note that
Tobler, Getis, Gould, and other researchers are presently experimenting with hodological-space measures of the concepts of
distance and direction for possible use in geographic models.[18]

The critical feature of Lewin's work is, however, that he
argued that accumulated information about a system in which each
individual operates is a key factor influencing both locomotions in space, and the gap between perceived and physical
space.   In other words the ability of an organism to cope with
a space-system depends on what he can learn about it.
Obviously this statement has as much relevance for the geographer interested in behavior as it does for the psychologist
interested in the learning mechanism.

Place Learning

Tolman's learning theory argues that an organism learns not
by learning movement habits, but by learning the location of
paths or places.   In other words, he argues that learning is
a cognitive process guided by spatial relationships rather than
by reinforced movement sequences.[19]   This theory is known as
sign learning and it can be defined as an "acquired expectation that one stimulus will be followed by another provided
that a familiar behavior route is followed."[20]   The movement
that results is variable; i.e., one movement may be substituted
for another provided that both movements lead to the same
end point where a stimulus or reward is expected.   Habit formation theories (conditioning) assume that what the organism

learns are movements or responses to stimuli. Sign learning
proposes that under some circumstances the organism instead of
learning movement habits, learns the location of paths and
places!

Latent learning also supports the theory that spatial
orientation rather than sets of movements are learned. Latent
learning refers to any learning that is not demonstrated by
behavior at the time of learning.[21] Such learning occurs
under low levels of drive, or if drive is inactive, when
incentives to learn are lacking. When drive is heightened or
incentives appear, there is recall and use of what was learned.
For example, through "mild curiosity" we note the location of
a store selling goods in which we aren't at the moment interested. However, when we want something the store sells, we
head directly for it even though it was never before a goal-
object.

In association with this notion that mental maps influence
behavior, Tolman also investigated the problem of "reward expectancy" or levels of aspiration. When a shopping trip for
(say) food buying is undertaken, then obviously there are
expectations concerning the type and quality of food that is
sought. If a reward is merely something that satisfies a
drive, one type of reward (e.g., food, clothing) ought to be
as useful as another type (i.e., food, clothing) as long as it
is generally regarded as "acceptable." But if a certain kind
of reward is expected, alternatives which were not expected
may be rejected. For example many people shop for specific
brands of goods. If at a given store the expected brand is
not available then the consumer may go elsewhere -- even
though an adequate array of generally acceptable substitute
brands are available! Thus multiple-place shopping may occur
when a given level of aspiration is not achieved, even though
close substitutes for a good or service are available. This

concept has been implicitly used in geography to account for
unexplained variation in shopping-behavior models.

Contiguity

Geographers have traditionally explained the location and
distribution of phenomena by looking for variables areally
associated with or contiguous to the problem variable.
Guthrie's learning theory should be inherently appealing to
geographers because it similarly argues that the influence of
a rewarding state of affairs acts not only on a given stimulus-
response connection, but also on acts in the neighborhood of a
rewarded one.[22]   Thus there is a spread effect in learning
which reinforces responses in the vicinity of (or which are
associated with) rewarded response.

Guthrie's theory of learning is also based on the convic-
tion that behavior can be studied only when it is overtly
observable.   Very frequently overt acts have some spatial
manifestation -- and these spatially overt acts are precisely
the behavior that geographers try to explain.   This is then a
theory concerned with movements themselves rather than whether
movements lead to success or error.

One further aspect of this theory that has led to its
mathematical interpretation, is the argument that "the animal
learns to escape with its first escape...."[23]   This has led to
the interpretation of learning as a Bernoulli-type process and
ultimately to the representation of learning as a Markov process.[24]   He also called attention to repetition and stereotypy in behavior after learning was achieved and of course
this is the type of behavior geographers frequently seek to
describe and predict.   Another interesting conclusion from his
work was that rewards do not strengthen behavior but only prevent it from disintegrating.   Although little research into
extinction has been undertaken in geography, it also promises
to be a fertile field of inquiry.

Habit

The central concept of Hull's theory is habit.[25] Most of
his information about habit formation is derived from experiments with conditioned responses (reinforcement theory).
A major thesis of this theory is that rather than contribute
its maximum influence on one trial reinforcement adds an increment to habit strength on each occurrence. Habit strength is
defined as a positive growth function of the number of trials,
and is represented by the formula:[26]

$$_sH_r = 1-10^{-aN}$$

when $_sH_r$ = habit strength   (1)

N = the number of trials

a = an empirically defined constant
        (.03 in Hull's work).

In addition to conditioning, Hull examined such phenomena as
trial and error behavior, discriminatory learning, maze learning, and other useful concepts. The end result of this research was an incremental learning theory.

The theory that learning is a gradual process includes
within it the notion that behavior changes from motivated
search (trial and error) to fully learned activities. Since
this transition involves a variety of spatial manifestations,
it is pertinent to examine in detail some of its concepts.
For example, many problems require the selection of one or
another mode of action to reach a goal. Often alternative
responses appear successively in an unorderly fashion until a
satisfactory response is made. Such procedures are called
trial and error procedures and their basic element is search
activity. If an individual is placed in an unfamiliar environment and stimulated to seek a goal, he generally exhibits
a tendency to vary his responses (law of multiple responses).
These responses are made under conditions of uncertainty as to

their outcome. After a range of trials, the "correct" or
most satisfying response is retained (the law of effect).[27]

In search practices, prior experiences will influence the
type of search patterns -- this is particularly true if there
is a transfer of knowledge from previously experienced search
systems. The first stage of a search procedure is called the
"provisional try." This is an attempt to find a satisfactory
response pattern and is sometimes referred to as hypothesis
behavior. The provisional try achieves an outcome and is
corrected by the feedback of information concerning the consequences of the try. Thus, a try that is rewarded is favorably reinforced and may become the basis for the formation of
an habitual reponse. A try that is unsuccessful or reinforced by punishment may be deleted from the response pattern
and an alternative search pattern substituted for it. Geographically we recognize some regularities in the search process. For example, we often adopt an assumption that search
will take the form of trying first the closest alternatives
and trying last the furthest alternatives (least effort hypothesis). Alternatively, we often argue that there is a definite trade-off between expected satisfaction from a trip and
distance, such that in multiple response situations there develop regular "decay functions" which specify probabilities
that search will be extended over certain distances. In fact,
both these assumptions form the basis of our most powerful
urban theories.

The principal idea to be retained at this stage is that
while search may be a random procedure more likely it is undertaken according to a definite set of rules. In fact, it can be
argued that once beyond the first sensorimotor experiments,
an individual's responses will never be purely random. We
can even postulate at this stage that stereotyped search
procedures may develop in which outcomes are samples

systematically according to, say, a locational or directional
bias. The degree of success resulting from this type of
searching may govern the degree to which an approach is made
toward a rational or maximizing or habitual behavioral pattern.

Hull also recognized that there are multiple response
paths between any origin and goal, and consequently individuals
learn alternative ways of traversing routes. These alternatives form a habit "family" which are arranged in a preferred
order.[28] For example, short routes may be more strongly preferred to long routes. Less favored routes are chosen when
the more favored are blocked in some way. Types of spatial
barriers and their effect on route selection have been defined
by Yuill[29] and by researchers interested in diffusion models.[30]
The habit-family concept is useful for examining the problem
of the attraction of goals behind barriers and the selection
of paths around barriers. It is worthwhile noting here that
many geographical models of spatial behavior have used an
assumption of free space (no barriers) and often modifications
of theory involve introducing barriers and examining resulting
changes in movement systems.

The Concept of Stereotyped Responses

The principal characteristics of a stereotyped response
are rigidity (invariability), repetition, and resistance to
change (persistence). Stereotyped actions consist of constantly elicited patterned responses which have a high degree
of habit strength.[31]

Most geographical theories which include behavioral elements and produce optimizing predictions rely implicitly on
the assumption of stereotyped behavior. In many cases the
"stereotyped" assumption is necessary in order to have any
theory at all. Traditional city-hinterland theories for
example assume the existence of a least-effort syndrome which
produces a stereotyped action of always patronizing the closest

occurrence of a phenomenon. Unfortunately, the stereotyped
action assumed in this case is a trivial one -- the repeated
patronage of a single node.   In practice while recognizing
that stereotyped actions do occur, most of the actions involved
in various types of marketing behavior are of the multiple response type. This is true both of shopping trips within a
city and patronage of urban centers by the dispersed population.

One of the most significant factors related to the existence of spatial stereotyped action is that its occurrence
makes possible the formulation of spatial behavioral theories.
We can in fact regard stereotyped responses as asymptotes of
a learning process; once this stage is reached then it becomes a simple matter to predict future responses.   However,
we do not know how long it takes an individual or a group to
achieve a stereotyped phase, nor do we know what proportion of
any given group at any time has already achieved stereotypy
and how many are still in a search phase.   This means that
the degree of successful prediction by models which include
stereotyped action as a necessary condition may be very small.
In fact, this is the situation experienced when attempts are
made to apply existing geographical models of consumer behavior to real world conditions and deduce from them the locational
patterns of some of man's activities. Despite insufficient
empirical findings about stereotypy, we can suggest as a hypothesis for the time being, that stereotyped actions in space
develop as part of a process that aids the organism in adapting to its environment by providing it with a relatively
organized and systematized conception of the multiple and
everchanging events and experiences of life.   Without such
a conceptualization, it would be very difficult to find any
regularity in the actions and behavior of humans.

Choice and the Decision Process

An essential element of the learning process is the choice of responses. Some theories describe learning as a change in response probability as the result of correct and incorrect choices in a continuing decision process.[32] Within the field of geography, Huff, Thompson, and Wolpert have discussed various aspects of the decision process and corresponding choice behavior.[33] For example, Huff has attempted to conceptualize the transition from a premotivated stage of behavior to the overt spatial act. This disregards quiescent or unmotivated behavior (such as neural or synaptic responses) which are of little concern to the geographer, but it also neglects the recursiveness associated with the assessment of an overt act due to levels of aspiration and the activities of learning. In short, once an overt spatial response has been made, individuals will probably restructure their decision processes in the light of accumulated information (i.e., after the basic unit of learning has been achieved). Either the provisional try behavior will be repeated or a new response choice will be made. Eventually a firmly established response pattern will emerge which is regularly triggered by presentation of the original stimulus condition. The asymptotic result of a repetitive choice process is the formation of stable-state choice proportions, or strategies, which govern future behavior.

The examination of choice processes also provides information on the idea of "rational behavior." In the past geographers and other researchers have considered "rational" and "economic maximization" to be synonymous. Siegel, on the other hand, shows that both pure and mixed (matchings) strategies can be considered "rational."[34] He equates economic rationality with the development of pure strategies, but also argues that if boredom, curiosity, and other motivations are considered, economic gain may not dominate behavior-strategies.

114

In fact he constructs a model of the choice process with two critical choice components -- the utility of correct choice, and the utility of variability.   It would appear that geographers can make considerable use of both concepts in their own research on choice behavior.

It is apparent from the foregoing discussion that a number of learning theories and many of their key concepts have definite spatial implications.   The possibility of using such theories in geography is enhanced by their translation into objective mathematical models which can be applied both to spatial and nonspatial situations.   The following sections provide a review of some recent learning models together with suggestions for their spatial applications.

C. Recent Learning Models with Possible Spatial Applications

The recent attention to the representation of learning through the media of mathematical models has led to a refinement and operationalizing of many elements of learning theories. Considerable attention in current research is being focused on models such as:

a)   Association models; examples include concept-identification models, paired associate models, and linear operator models;

b)   stimulus sampling models, based either on linear difference equations or stochastic processes;

c)   interactance - process models, based on either game theory or Markov models;   and

d)   avoidance conditioning models using linear difference equations.

Most current learning models are probabilistic in nature. This implies that some variability of behavior is built into the models.   By interpreting behavior as a stochastic process, for example, it is argued either that it is intrinsically

probabilistic, or that it is determined by antecedent conditions which encourage variability from one time period to the
next.

Geography is sometimes characterized as being in a phase of
development in which it is still trying to discover how to
measure its major dependent variables. To some extent we can
claim affinity with psychologists such as Guthrie, Skinner,
and Estes who have argued that response probability is their
most appropriate dependent variable. Certainly there are many
spatially overt acts which we measure in terms of "response
probability" -- i.e., frequency of trip-making; frequency of
contact, etc. -- and which we can classify into different
types of response classes. Thus, models which have as inputs
spatial variables, and which illustrate changes in response
probabilities over time appear suitable for both geographic
and psychological uses.

The Concept Identification Model

Perhaps the most elementary learning model that has potential use for geographers is the "concept identification" or
"discrimination learning" model. This is presented either
from the stimulus-response reinforcement viewpoint of Hull,
Spence, Burk and Estes,[35] or as a model of hypothesis behavior,
as in the work of Restle or Levine.[36] The former viewpoint
argues that subjects discriminate among responses according to
type, force and frequency of reinforcement and that the end
product is habit formation. The hypothesis viewpoint argues
that responses are elicited by a set of stimuli after trial
and error behavior has occurred. Such trial and error behavior is essentially a sequential testing of hypothetical responses, with the aim of finding a satisfactory stimulus-
response pairing. The discrimination learning model is based
on the task of finding which cue is relevant in a given problem
situation. Once the cue is recognized, future responses are

based on this. In other words the individual has learned a
response pattern and may develop a habitual response to it.
Market researchers have used these concepts to show the value
of labels and advertising in the selection of goods.[37] Geographers also can examine analogous conditions. For example,
in the development of a journey-to-work pattern, we presume
that travelers select alternative routes from a total array of
routes and then test them. If a choice (cue-identification)
leads to an error or unsatisfactory trip sequence, that route
may be discarded and a new choice made. Cues that influence
selection of routes have so far been identified as least
effort, shortest time, maximal aesthetics, fewest barriers,
least boring route, and so on.

Despite the large number of criteria that may influence
choice of a response path, generally there are cues that are
more obvious than others. For example, for high order
(shopping) goods the most relevant cue for achieving a satisfactory response might be one of the following: presence of
a department store in the center chosen for shopping; a
minimum number of functions in the chosen center; or distance
to be traveled. Irrelevant cues might be: whether or not a
friend recommends the center; presence of a cafeteria;
presence of a movie theater; or presence of <25 functions.

Once a subject learns which cue is most relevant, he bases
future actions on this. For example, Berry has argued that
the number of functions is a relevant cue for predicting probabilities of trips to urban centers.[38] Golledge, Rushton
and Clark argue that a maximum distance is the most relevant
cue.[39] A study recently conducted in Columbus, Ohio, by the
author resulted in identification and ranking of cues in the
order given in Table I. Support for both the above hypotheses
is evidenced by the data collected in this study. One conclusion drawn from this study was that as we alter the trip

purpose, the relevant cues change, but once a relevant cue has been identified for a specific purpose, it influences future behavior.

To translate these theoretical notions into a model, we must represent mathematically the "state of knowledge" of a subject with respect to the solution to a problem. Therefore, let us:

a)   denote by S a solution state where a subject
     holds a "correct" relevant cue as his hypothesis
     before a test;

b)   denote by $\bar{S}$ a presolution state where irrelevant
     cues are followed so that hypothesis testing
     will result in unrewarded (incorrect) responses.

Let the probability of a correct response when in $\bar{S}$ be p. Assume the subject starts in state $\bar{S}$ on the 1st trial. "Learning" is said to have occurred when the subject makes a transition from $\bar{S} \longrightarrow S$ on one trial and thereafter remains in S.   (i.e., when S becomes an absorbing state).   Let probability of an error when in $\bar{S}$ be q = (1-p), and the probability of a relevant cue being selected following an error trial be c.   Then the probability of the joint event "make an error and select relevant cue" is qc.   Thus qc is the probability on any trial that a subject who started the trial in state $\bar{S}$ begins the next in S.   The probability that a transition $\bar{S} \longrightarrow S$ fails to occur is: the likelihood (p) of a correct response in $\bar{S}$, plus the joint likelihood of an error and failure to select the relevant cue:

$$\text{i.e.} \quad p + q\,(1-c) = 1-qc \;. \tag{2}$$

Thus (1-qc) is the probability that a subject starting in state $\bar{S}$ is still in that state at the next trial.

If we take a sample of responses we generate a sequence of the states $\bar{S}$ and S.   The model aims at finding the generation rule of the probability process generating the outcomes (or sequence of $\bar{S}$ and S) for each subject.

TABLE I

Cues for Selection of Shopping Centers

Attribute	Frequency of Rank I	Proportion of Times Attribute was ranked # 1
Closeness	112	35.7
Variety of Stores	90	38.7
Quality of Products	32	10.2
Parking	28	8.9
Prices	15	4.8
Service Quality	6	2.0
Freeway Access	4	1.7
.	.	.
.	.	.
.	.	.

Source: Shopping Center Utility Survey,
Department of Geography,
The Ohio State University, 1967.

Some outcomes might be:

$$\bar{S} \quad S \quad \bar{S} \quad \bar{S} \quad \bar{S} \quad S \quad S \quad S \quad S \quad . \quad . \quad . \quad .$$
$$S \quad \bar{S} \quad \bar{S} \quad S \quad \bar{S} \quad S \quad S \quad \bar{S} \quad \bar{S} \quad S \quad S \quad S \quad . \quad . \quad . \quad .$$
$$S \quad S \quad \bar{S} \quad \bar{S} \quad \bar{S} \quad S \quad \bar{S} \quad S \quad S \quad S \quad S \quad . \quad . \quad . \quad . \text{ etc.}$$

This is a sample space of outcomes each having a probability p_i of occurrence.

An event on this sample space is a set of outcomes, e.g., "process is in state $\bar{S}$ on the 5th trial" is event $\bar{S}_S$. The probability of an event is the sum of the probabilities of sample points that are members of the set defined by event $\bar{S}_S$, or

$$P_S = \frac{\text{\# outcomes in which } \bar{S}_S \text{ occurs}}{\text{total \# of outcomes}}$$

It should now be obvious that the discrimination learning model can be formulated in terms of a Markov chain. Given that either $\bar{S}$ os S occur on trial n, the Transition Probability Matrix (TPM) is:

$$
\begin{array}{cc}
 & \begin{array}{cc} S_{n+1} & \bar{S}_{n+1} \end{array} \\
\begin{array}{c} S_n \\ \bar{S}_n \end{array} & \left[\begin{array}{cc} 1 & 0 \\ qc & 1-qc \end{array} \right]
\end{array}
$$

Here the 1 in a_{11} identifies S as an absorbing state. Given the sequence:

$$\bar{S}_1, \; \bar{S}_2, \; \bar{S}_3, \; S_4, \; S_5, \; - \; - \; - \; - \; . \; .$$

the appropriate transition probabilities are:

$$1 \; . \; (1-qc) \; . \; (1-qc) \; . \; qc \; . \; 1 \; . \; 1 \; . \; 1 \; . \; . \; . \; . \; . \; . \; = (1-qc)^2 qc$$

If we let Y be a random variable equalling the number of trials the process is in state $\bar{S}$ before shifting to S, then the above sequence lets Y = 3, and

$$Pr(Y=3) = Pr(\bar{S}_1, \; \bar{S}_2, \; \bar{S}_3, \; S_4, \; S_5, \; . \; . \; . \; .) = (1-qc)^2 qc.$$

The probability distribution for this random variable Y is:

$$Pr(Y=n) = (1-qc)^{n-1} qc. \tag{3}$$

The discrimination-learning model therefore can lead us to represent the learning process as a 2-state absorbing Markov chain. As such, it implies that the sequence of presolution trials constitutes a Bernoulli series of observations with mathematical properties of stationarity and statistically independent outcomes.

Applications of this model to spatial problems may not be immediately obvious. Much of the usefulness of the model probably lies in the interpretation of its output and its use in sample selection, rather than in examining input data or model structure. An example will help to illustrate this point.

120

Assume that, in a given city, we are interested in the patronage of shopping centers. From existing empirical studies we know that proximity, number of functions, and the order of the highest ranked function appear to be significant variables in shopping center selection. However, combining these variables into a multivariate model (e.g. a multiple regression equation), may give a low level of explained variation for a given population. To check the validity of the study, and to help explain this low level of explanation, a concept identification experiment could be carried out. This would involve giving sample members a list of relevant shopping centers and their attributes, and asking respondents to predict the center in which they could expect to buy a good (e.g. expensive jewelry). For each center in turn a Yes-No answer is given by the respondent, and this is immediately reinforced by a Correct/Incorrect response from the interviewer. Trials are repeated with a mixing of the order of presentation, until relevant cues for completing a purchase are recognized (i.e. learned). An experiment of this type was actually conducted for a small population (N=24) to test the usefulness of the technique.[40]

The data obtained from the above experiment was analyzed by a concept identification model of the form:

$$q_n = q(1-qc)^{n-1}qc \quad \text{(q,c, defined as before).}$$

The estimation of the parameters q and c proceeds as follows:

$$\hat{p} = (1-q) = \frac{Z}{Z+T}1 = \frac{Z}{L} \tag{4}$$

where Z = total number of correct responses before the last error for all individuals.

T^1 = total number of errors for all individuals. The results of the experiment included estimates of p, q and c, and the derivation of some useful statistics from the basic model.

e.g.

$$\hat{p} = \frac{409}{409 + 126} = .76$$

$$\hat{q} = 1 - \hat{p} = .24$$

$$\hat{c} = \frac{N_1}{T} \quad , \text{ where } N = \text{ the number of subjects}$$

$$= \frac{24}{126} = .19$$

Statistics that can be derived from the error counts are:

(a) $E(T) = $ mean value for $T = \frac{\Sigma T}{N} = \frac{126}{24}$

$\quad\quad = 5.25$; this represents the mean total number of errors before learning.

and

(b) $(\text{VAR } T) = E(T^2) - E(T)^2 = 38.58 - 27.56 = 11.02.$

Let $P(T=k)$ be the probability that a person will make k-errors before learning. This is generally defined as:

$P(T=k) = c(1-c)^{k-1}$. All this information can then be used to assess and stratify the sample population.

Each of the statistics derived above provide interesting information for the geographic researcher. For example it was earlier suggested that a random sample of urban dwellers for a consumer behavior study would probably include all the stages from search to stereotyped behavior. Such a representation may have produced the low level of explanation in the original regression model. Information given by a preliminary study such as this one could provide the basis for stratifying the sample population. If the mean number of errors for any given task is known, then extremely variable respondents (i.e. those who do not rely on the cues which are to be incorporated into an explanatory model) can be deleted from the study. The researchers could also ensure

that all their sample population had undertaken at least E(T)
shopping trips to the relevant centers. The result is improved prediction, but more important still, it shows the proportion of the population for which the explanatory model is
an "adequate" analytical tool.

Paired Associate Models

Paired associate learning models appear to hold considerable
promise for the geographer. In this case there is probably
equally as much potential in the way data is collected and
prepared as there is in the models using such data.

In this type of learning experiment subjects are given a
list by pairs of stimuli (e.g. nonsense syllables) and asked
to learn the list. While this type of experiment appears
fruitless for the geographer, the methodology is not. For
example Rushton[41] has interpreted space-preference as a choice
between pairs of alternatives and this has allowed him to use
powerful multidimensional scaling techniques to determine the
choice of urban centers by a dispersed population. Similarly,
by regarding the selection of shopping centers as a paired-
comparison procedure, Briggs[42] has defined a utility scale for
the selection of shopping centers by a sample population.
Further developments in this area are obvious and promising.

Although the multidimensional scaling techniques can be
used to illustrate learning by showing changes in the configurations outputed at various time intervals, this line of
research is a fairly recent one and there is insufficient information to provide a summary of its relative worth. There
are however many other more conventional models used in paired-
comparison experiments, including one-element models,[43] single-
operator linear models[44] and random trial increments models.[45]
For purposes of convenience only the single-operator linear
model will be examined here.

The Single Operator Linear Model views learning as a direct change in response probability from one trial to the next.[46] Thus response probability on trial (n+1) is obtained by a simple transformation of response probability on trial n. If q_n is the error probability on trial n, then this model generates a sequence of probabilities $(q_1, q_2, q_3, \ldots)$. It assumes that the transformation of $q_n \longrightarrow q_{n+1}$ is linear. In the single operator linear model, the same operator is applied to q_n on every trial. The model is mathematically a first-order difference equation. Its general form is:

$$q_{n+1} = \alpha\, q_n \qquad (5)$$

where α = a fraction representing the extent to which error probability on trial n is reduced for trial (n+1) as a result of reinforcement, and q_{n+1} = a linear function of q_n. To illustrate the basic difference equation:

as $P_{n+1} = 1 - q_{n+1}$

and since $q_{n+1} = \alpha\, q_n$

and $q_n = 1 - P_n$

then $P_{n+1} = 1 - \alpha\, (1 - P_n)$

$\qquad = \alpha\, P_n + (1-\alpha)$

or: $P_{n+1} = pn + (1-\alpha)\,(1-P_n) \qquad (6)$

That is, the probability of a correct response on trial (n+1) is the sum of the prior value of P_n and an increment which is a proportion of the maximum possible gain $(1-P_n)$.

Since this type of model has already been investigated by geographers, no examples will be developed here.[47] However it is relevant to suggest some spatial characteristics of the model.

The literature of geography abounds with market area studies. One of the main conclusions drawn from these studies is that, as distance from a given center increases, the likelihood of patronizing it diminishes. In other words, decreasing probabilities of patronage can be partly explained by a decrease in the probability of learning about the advantages of a center as distance increases. Alternatively it can be argued that as distance changes, the likelihood of approaching an asymptotic pattern limited to single center patronage, is correspondingly diminished. In terms of the linear operator learning model, both the λ_i (asymptotic values) and α_i (learning parameters) may vary over distance. At present no evidence exists to show the manner of this variation, but the calculation of this change through space presents a challenge to the geographer.

It can also be expected that statistics derived from the model will vary consistently with distance, so that mean total "errors" will increase as will the number of "trials" before any asymptote is reached. The relative advantage of this information for different classes of goods (and therefore different sizes of center) is obvious and will not be further developed.

Interactance-Process Models

One of the most widely occurring problems in the social and behavioral sciences is the problem of predicting the outcome of a choice process when two or more alternatives are placed in competition with one another. A variety of models have been derived to handle choice decisions, ranging from locational-choice models in geography,[48] to resource-allocation models in economics,[49] and preference-ordering models in psychology.[50] The complexity of such models varies from simple gravity models used to predict the choice of market centers, to complex gaming models with mixed strategies. The versatility of many choice models allows them to be used both to describe the choice process and to describe the resulting interactions that occur.

Descriptions of the choice-act generally include the following information:

a. specifying the alternative eventually chosen
   (a posteriori choice)

b. specifying the nature of any trial and error
   behavior that occurs

c. specifying the time taken to respond

d. an expression of confidence in the correctness
   of choice, and

e. a statement of the difficulty experienced in
   making choice.

Geographers have most frequently dealt with a posteriori choice and have neglected other aspects of the process. This is largely because overt spatial activity results from the choice act, and it is this activity that most interests the geographer. The data recorded by geographers is simply the alternatives selected, and then spatial and other variables are used to explain why the specific alternatives were chosen. Sometimes possible response strength (or "potential") is estimated from a series of overt spatial acts, and potential-type models are used to predict future interactions. It is interesting to note that recent geographic research has gone beyond analysis of a posteriori choice to examination of the antecedent and concurrent conditions influencing choice.[51] It is for this type of research that variations of learning models become relevant, and assumptions such as the weak stochastic transitivity assumption, Luce's choice axiom, the constant ratio rule, and the scaling theories of Torgerson and Coombs become significant.[52] Researchers contributing models of this type include Hull, Spence, Siegel, Simon, Atkinson, Suppes, Kruskall, and numerous others.[53]

Luce's Choice Axiom concerns itself with the relationship between choice probabilities as the number of alternatives

involved in the choice act change.[54]   Its basic premise is
that the ratio of the likelihood of choosing element (a) to
the likelihood of choosing element (b) in a set of k-alternatives is a constant irrespective of the number and composition
of the other alternatives in the choice set.

Let: x, y, z, t, u, ... be alternative elements

   T be the total set of alternatives

   R be some subset of T

   $Pr(x;R)$ be the probability that x is chosen
      when the choice is restricted to subset R

   $Pr(x,y)$ be the probability that x is chosen
      from a subset of x and y.

Then $Pr(x;T) = Pr(R;T).Pr(x;R)$.   (7)

If we choose an arbitrary element (a) of T, then the response
strength of x, $(v(x))$ is:

$$v(x) = \frac{Pr(x,a)}{Pr(a,x)} = \frac{Pr(x;T)}{Pr(a;T)}$$

Response strengths of elements have the following property:

$$\frac{v(x)}{v(y)} = \frac{Pr(x,y)}{Pr(y,x)}$$

If we let $Pr(x,T) = \dfrac{1}{\sum\limits_{y} \dfrac{Pr(y,x)}{Pr(x,y)}}$ (8)

then $Pr(x:T) = \dfrac{1}{\sum\limits_{y} \dfrac{v(y)}{v(x)}} = \dfrac{v(x)}{\sum\limits_{y} v(y)}$ (8a)

Note also that if these response strengths (or "scale values")
are multiplied by a positive constant, the original relationship of scale values to choice probabilities is retained.
In order to calculate probabilities of choice either in pairwise choice situations or where k-alternatives are ranked, it
is necessary to define a ratio w_{xy}:

$$w_{xy} = \frac{Pr(x,y)}{Pr(y,x)} \qquad (9)$$

Thus in a pairwise choice experiment for 3 objects

$$\frac{Pr(x,y)}{Pr(y,x)} \cdot \frac{Pr(y,z)}{Pr(z,y)} = \frac{Pr(x,z)}{Pr(z,x)}$$

$$\frac{Pr(x,z)}{Pr(z,x)} = \frac{Pr(x,z)}{1-Pr(x,z)} = w_{xz}$$

$$\therefore \quad Pr(x,z) = w_{xz} - w_{xz} \cdot Pr(x,z)$$

$$= \frac{w_{xz}}{1 + w_{xz}}$$

Once estimates have been made of any given w-ratio, estimates of the probability of choosing one element over another can be made. For example a sample of potential or actual shopping center patrons are presented with a list of 3 alternative centers and, using paired comparison procedures, are asked for their choices between the centers.

Results may be:

TABLE 2

Paired Comparison of Shopping Centers

		1	2	3
	1	-	0.60	0.90
Shopping Center	2	0.40	-	0.80
	3	0.10	0.20	-

Source: Hypothetical data.

If we were given results for $Pr(1,2) = 0.60$ and $Pr(2,3) = 0.80$, we could estimate $Pr(1,3)$ from the choice axiom as follows:

$$w_{12} \cdot w_{23} = \frac{Pr(1,2)}{Pr(2,1)} \cdot \frac{Pr(2,3)}{Pr(3,2)}$$

$$= \frac{0.60}{0.40} \cdot \frac{0.80}{0.20} = 6 = w_{13}$$

and $\qquad Pr(1,3) - \dfrac{w_{13}}{1 + w_{13}} = \dfrac{6}{7} = 0.86$

Despite the apparent promise of this choice model, empirical testing of its dominant thesis in a spatial framework has not yet produced good results.[55] Apparently the reason for this is that it is useful for repetitive choice situations for individuals, but cannot easily be aggregated so that conclusions can be drawn for group behavior, unless it is assumed that all individuals in the group had the same basis for stating preferences. With a population of individuals whose locations vary, this limiting assumption is not feasible.

A more general use of Luce's work can be found in the realm of stimulus ranking. In ranking experiments which are not conducted by pairwise choice procedures, stimulus ranking theories, using the choice axiom, allow us to predict the probability with which each rank occurs. In this case, the rate of change in response strengths from trial to trial is interpreted as the learning parameter.

The Stimulus Ranking Model also involves calculating probabilities for ranking experiments. Ranking of a set of k alternatives amounts to advancing a hypothesis about the way people generate and order their preferences. It is possible to use Stimulus Ranking theory to predict the probability with which each ranking will occur, and the relative order of the rankings. By itself this amounts only to a cross-sectional

description of the order of choices. When combined with
choice theory and placed in a stochastic framework however, it
is possible to account for features of choice behavior in
addition to specifying the alternative chosen; for example,
the trial and error behavior prior to final decision-making,
the time taken to arrive at a fixed choice pattern, the degree
of confidence in the decision, and so on, can all be included
in the one model.

The first prominent theory of this type was Tolman's.
Recently Estes, Spence, Atkinson and LaBerge have used the
concept of random walks with absorbing barriers to operationalize and expand Tolman's hypotheses.[56] For example, given
the following choice situation.

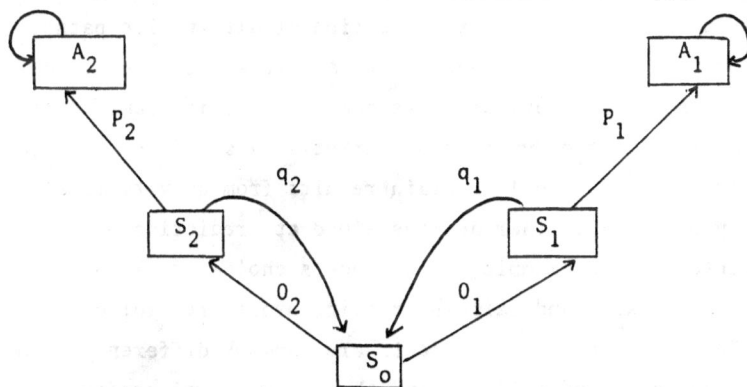

where S_0 = the choice point -- a neutral state

0_i = probability that a subject orients to S_i when
at the choice point: $(0_1 + 0_2) = 1$.

P_i = probability that end point A_i is approached
and the trial is terminated.

A_i = an absorbing state representing achieving a goal.

q_i = probability of orienting away from S_i and
reverting to the neutral state S_0.

The model describing the probability of absorption at, say, A_i is:

$$Pr\ (A_1) = \sum_{n=0}^{\infty} {}_0{}_1p_1\ ({}_0{}_1q_1 + {}_0{}_2q_2)^n \qquad (10)$$

Essentially, finding the probabilities of absorption involves evaluating the sum of a geometric series with ratio $({}_0{}_1q_1 + {}_0{}_2q_2)$. In the above example, the general term for n loops before absorption at A_1 is $({}_0{}_1q_1 + {}_0{}_2q_2)^n{}_0{}_1p_1$. From this model can be derived the probability distribution of the number of loops before absorption (L), which is again geometrically distributed. One interesting conclusion from this type of model is that when stimuli are of low attractiveness, there will be a large amount of trial and error behavior before absorption. This means that by examining the behavior of individuals with respect to selection of alternative paths, or alternative goals, we can estimate the relative strength of the goals. This obviously has some interesting ramifications for studies of the dynamics of market-area analysis, and could perhaps be used to help explain results from gravity models, Reilly's law, and other devices aimed at predicting patronage of centers. An example, using Luce's choice model and both stimulus ranking and pairwise ranking procedures follows.

In this experiment, subjects are shown k different stimuli simultaneously and told to rank them in order of preference. Luce assumes the individual picks the most preferred object and ranks it 1, then considers the remaining k-1 objects for rank 2, etc. Thus a subject makes k-1 successive decisions. He also assumes that each choice depends on the ratio of the scale values, computed from the stimuli remaining in the reduced set. e.g. Assume we have three objects A, B, C, with v-scale values a, b, c. There are 3! or 6 possible rankings:- ABC, ACB, BAC, BCA, CAB, CBA. With replication of the experiment, each of these will occur a certain proportion of

time.   Luce aims for a choice theory to predict the probability with which each ranking occurs, and quantities such as
the probability that A gets ranked second, that B's rank exceeds
A's, and so on.

Consider Pr(BAC).   From Luce's choice axiom we calculate
the probability that B is chosen out of 3 elements.   Then
eliminating B, we calculate the probability that A is chosen
over C in the 2-object comparison.   The joint probability of
the ranking BAC is the product of these 2 probabilities:

i.e. $\Pr(BAC) = \left[\dfrac{b}{a+b+c} \right] \cdot \left[\dfrac{a}{a+c} \right]$ (11)

Note here that the likelihood of an element of the total
set receiving rank (1) is the same as its likelihood of being
chosen from a set of k-alternatives when only one choice is
permitted.   For example:

$$\Pr(B=R1) = \Pr(BAC) + \Pr(BCA)$$
$$= \left[\frac{b}{a+b+c} \right] \cdot \left[\frac{a}{a+c} \right] + \left[\frac{b}{a+b+c} \right] \cdot \left[\frac{c}{a+c} \right]$$
$$= \frac{b}{a+b+c} \qquad (12)$$

Note also that there is a relation between the ranking of
k-alternatives and the pairwise choice data.   If the ranking
hypothesis is correct, then the likelihood of A being chosen
over B in a pairwise choice situation, equals the sum of the
probabilities of all cases where A receives a higher rank than
B in the k-alternative situation.

i.e. $\Pr(A,B) = \Pr(ABC) + \Pr(ACB) + \Pr(CAB)$ (13)

Since $\Pr(ABC) + \Pr(ACB) = \dfrac{a}{a+b+c}$

Then $Pr(A,B) = \dfrac{a}{a+b+c} + \left[\dfrac{c}{a+b+c} \cdot \dfrac{a}{a+b} \right]$

$= \dfrac{a(a+b)+ac}{(a+b+c)\ (a+b)} = \dfrac{a^2+ab+ac}{(a+b+c)\ (a+b)}$

$= \dfrac{a(a+b+c)}{(a+b)\ (a+b+c)}$

$= \dfrac{a}{(a+b)}$

Next, we can define the probability that an arbitrary element (C) receives a given rank (2) when the elements are ranked. This involves summing the probability of all cases when C gets a rank of 2.

$$Pr(C=2) = Pr(ACB) + Pr(BCA) \tag{14}$$

$$= \left[\dfrac{a}{a+b+c} \cdot \dfrac{c}{c+b} \right] + \left[\dfrac{b}{a+b+c} \cdot \dfrac{c}{a+c} \right]$$

$$= Pr(A=R1).\quad Pr(C,B) + Pr(B=R1).\quad Pr(C,A$$

Assume for the sake of an example, we ask a sample of people to rank three shopping centers -- which for convenience we shall call Northland, Westland, and Eastland. Then Table 3 summarizes the rankings of the population.

As another possible alternative, let us assume that we have the pairwise rankings of these centers provided by the same population (see Table 4).

TABLE 3

Stimulus Ranking of Choices

Shopping Center	Rank			Mean Rank
	1	2	3	
Northland (N)	0.65	0.32	0.03	1.38
Eastland (E)	0.29	0.43	0.28	1.99
Westland (W)	0.06	0.24	0.70	2.64
Sum	1.00	0.99	1.01	

TABLE 4

Pairwise Ranking of Choices

	Northland	Eastland	Westland
Northland	-	0.71	0.89
Eastland	0.29	-	0.73
Westland	0.11	0.27	-

(NB.   These are hypothetical figures only.   Note that entrances on the diagonal would represent comparing the centers with themselves.   This comparison is not generally made unless one is interested in time-order or space-order errors).

To predict the numbers in the stimulus ranking table from Luce's ranking hypothesis, consider first the probability that N receives rank one:

$$Pr(N=R1) \; = \; \frac{n}{n+e+w} \; = \; \frac{1}{1+\frac{e}{n}+\frac{w}{n}} \qquad (15)$$

From the table of pairwise choices:

$$\frac{e}{n} = \frac{Pr(E,N)}{Pr(N,E)} = \frac{0.29}{0.71} = 0.409$$

$$\frac{w}{n} = \frac{Pr(W,N)}{Pr(N,W)} = \frac{0.11}{0.89} = 0.124$$

Substitute into Eq. (15):

$$Pr(N=R1) = \frac{1}{1+0.409+0.124} = 0.652$$

Similarly:

$$Pr(E=R1) = \frac{e}{n+e+w} = \frac{\dfrac{e}{n}}{1+\dfrac{e}{n}+\dfrac{w}{n}} = \frac{0.409}{1.533}$$

$$= 0.267$$

$$Pr(W=R1) = 1 - 0.652 - 0.267 = 0.081$$

Now to predict the probabilities that each center receives a rank of 2.

$$Pr(E=R2) = Pr(N=R1). \, Pr(E,W) + Pr(W=R1). \, Pr(E,N)$$
$$= 0.652(0.73) + 0.081(0.29)$$
$$= 0.500$$
$$Pr(N=R2) = Pr(E=R1). \, Pr(N,W) + Pr(W=R1). \, (Pr(N,E))$$
$$= 0.267(0.89) + 0.081(0.71)$$
$$= 0.296$$
$$Pr(W=R2) = 1 - 0.296 - 0.500$$
$$= 0.204$$

This can be continued for the third rank probabilities, and final goodness of fit can be tested by a x^2- test.

The probabilities indicated above are obviously cross-sectional and apply only for a given period of time. Successive trials of this type of experiment would not only indicate the rate of learning (i.e. the rate of change of choice proportions) but would also give evidence of the feasibility of the single-center-choice hypothesis that is used frequently in existing geographic models. The relevance of this model and its output for geographers appears unquestionable.

Avoidance Conditioning Models

Avoidance conditioning models can be represented either by the two-element linear model of Bush and Mosteller or the non-linear β-models of Luce.[57] Both these models consider that learning takes place on each trial whether the outcome is reward (avoidance) or nonreward (shock). They also argue that reward and nonreward trials do not have equal effects and formulate models to account for these.

The general form of the linear model is

$$p_{n+1} = \begin{cases} \alpha_a \, p_n + (1-\alpha_a)\lambda_a & \text{if shock is avoided on trial n} \quad (16a) \\ \alpha_s \, p_n + (1-\alpha_s)\lambda_s & \text{if shock occurs on trial n.} \quad (16b) \end{cases}$$

where p_{n+1} = probability of successfully completing the (n+1)st trial (i.e., avoiding shock)

p_n = probability of avoidance on trial n

α_a = a fraction representing the extent to which success probability is increased on each trial when avoidance occurs

α_s = fraction representing the extent to which success probability is increased on each trial when shock occurs

λa = an asymptotic probability that could be reached as avoidance continues

λb = an asymptotic probability that could be reached if shock continues

In terms of the learning parameter (α), the bounds are $0 \le \alpha \le 1$. In this case the point where ($\alpha=1$) corresponds to the case where no learning occurs and ($\alpha=0$) corresponds to complete learning. Thus the smaller α is, the faster learning occurs. Values of α_i are usually estimated from tables provided by Bush and Mosteller.

This model has considerable potential for geographers and has been used in a one-element form by Golledge and Brown, and Haines.[58] Presently other tests of its applicability to spatial problems are being conducted at Ohio State University using both actual data and Monte Carlo simulation procedures.

For data that is seen to be best described by a nonlinear model, Luce's β-model may have some applicability. Luce formulated a nonlinear beta-learning model to describe changes in response strength from trial to trial. Essentially he argues that the theoretical effect of a learning trial is to multiply response strength by a constant. Since one outcome type (e.g., reward) may be more effective than another (non-reward), it is possible that the constants for each element in the choice set may differ. The β-index is defined in terms of a ratio of these constants. In this model, response strength is defined in terms of error probability (q_n):

$$\text{i.e., } v_n = \frac{1-q_n}{q_n}$$

The final format of the β-model is:

$$q_{n+1} = \frac{1}{\beta\left(\dfrac{1-q_n}{q_n}\right)+1} \qquad \frac{q_n}{q + \beta(1-q_n)} \qquad (17)$$

where q_{n+1} = error probability on trial $(n+1)$

$\quad q_n$ = error probability on trial (n)

$\quad \beta$ = ratio of the constants operating on each response strength.

Spatial Problems and Learning Models

Learning theories and their mathematical models describe changes in behavior from the first motivated act to a fully learned response or response sequence. The output from learning models can be used to describe all forms of spatial behavior from initial searching to repetitive (habitual) behavior. The entire learning sequence can be described by such models, or, alternatively, behavior at any point in time can be represented.

The versatility of learning models makes them potentially useful for geographers because the spatial actions with which we deal invariably encompass a range of behaviors from search to stereotyped habitual responses.

Frequently we have criticized optimal models of spatial behavior because they apparently do not fit "reality." Evidence for rejection is inevitably obtained by comparing optimal behavior with some empirical evidence of actions. Just as inevitably this empirical evidence is a conglomerate, composed of the actions of an unstratified group of people who may be at different stages of the learning process. Thus we have compared theoretical asymptotic behavior with a mix of search, partly learned, and fully learned behaviors, and as a result achieve a low level of explanation -- which leads to rejection of the model, reinterpretation of theory, and perhaps a frustrated researcher.

Obviously one way of overcoming this problem is to test models with cross-sectional data that is relevant and appropriate. An alternative is to undertake further research into behavioral processes and thus more fully understand the complex "reality" we often (rather naively) try to predict.

As pointed out by Simon, Katona, Wolpert, Gould, Golledge and Brown, and other interested researchers, few repetitive choice decisions are made without some preliminary search activity. One of the problems begging further research is the nature of space-searching activity. Gould[59] has offered a fine introduction to the problem, and I believe considerable progress could be made by using learning concepts and learning models.

However, not all spatial activity is search activity. Some
regular or habitual behavior pattern gradually evolves as a
coping strategy. Of course, as this type of behavior evolves,
the amount of space-searching should diminish. The formation
of habitual response patterns is of particular importance to
the geographer if these patterns involve stabilizing movements
in space, for this means that such movements are predictable
with a high degree of accuracy. The development of habitual
behaviors is important for studying journeys-to-work, shopping
behavior, recreational behavior, and in fact a variety of
social and economic interactions where repetitiveness is involved. Thus, examination of behavior of stratified groups of
immigrants and of long-time residents would precede the building of predictive models, and again the format of some learning
models appears appropriate for such an examination.

One of the consequences of migration within urban areas is
the extinction of some previously held responses and acquisition of new spatial habits. Both these problems can be studied
via learning models. However, the acquisition of new spatial
habits is not limited to recent migrants. Each time a new
urban function or a new response path develops, there is some
effect on existing behavior. Impact studies of this type can
be examined by both stimulus sampling and signal detection
models, and the output from such models can be used to simulate
future urban behavior.

Other problems which could benefit from study within a
learning-model framework include examining the role of pre-
trial information on trial behavior; finding the effect of
system "shocks" such as new avenues of transportation or new
"rewards" on spatial behavior; experimenting with extinction
processes as various types of spatial barriers intervene
between origins and destinations; and seeing if extinction of
a response in one segment of space inhibits interaction with
spatially associated goals.

Probably one of the most interesting and potentially useful outgrowths of experimentation with learning models comes
not so much from a particular model, but from manipulations
with pairwise-choice and ranked data. This leads the researcher to the field of metric and nonmetric multidimensional
scaling models, some of which are reviewed in Rushton's paper
in this volume. Use of such models would hopefully allow
examination of subjective utility or preference rankings and
should be particularly important to the field of consumer
behavior.

This list of models and problems is not comprehensive; it
only scratches the surface of a range of spatial problems that
could have some light thrown on them by using ideas and models
developed by learning theorists. The least impact of such a
move would be to encourage more frequent use of the dynamic
approach to behavioral problems, and help geographers move
away from cross -sectional analysis and the limited-use static
models that have served so well in the past.

Currently research is being undertaken on a variety of
spatial-behavioral problems. Interactance-process models
(i.e., choice models) are being frequently used,[60] and
attempts have been made to use linear-operator models (in mar-
ket-decision studies),[61] and some forms of signal detection
models (in diffusion studies).[62] It is hoped that continued
emphasis on spatial behavioral problems will lead to use of
some of the other type of learning models summarized in this
paper.

140

NOTES

1.   Such a definition does not limit behavior to humans,
but it does eliminate the actions of inanimate objects such as
rocks rolling, winds blowing, plants growing, etc.   It *does*
allow us to infer that behavior is *caused*, and has *directed-
ness, motivation, action, and achievement.*

2.   This process involves a number of identifiable steps
from making the first motivated response, through a series of
covert and overt behaviors until a basic unit of learning has
been achieved.   For a summary of definitions of learning see
E. Hilgard, *Theories of Learning* (New York: Appleton-Century-
Crofts, 1956, 2nd edition), 2-6.

3.   B.J.L. Berry, H. Barnum, and R.J. Tennant, "Retail
Location and Consumer Behavior," *Papers and Proceedings of the
Regional Science Association*, Vol. 8 (1962), 65-106;
S.H. Britt, *Consumer Behavior and Behavioral Science* (New York:
Wiley, 1966);   W. Isard and M. Dacey, "On the Projection of
Individual Behavior in Regional Analysis," *Journal of Regional
Science*, Vol. 4, No. 1 (Part I) and Vol. 4, No. 2 (Part II)
(1962), 1-35, 51-84;   E.N. Thomas, R.N. Mitchell, and
D.A. Blome, "The Spatial Behavior of a Dispersed Non-farm Population," *Papers and Proceedings of the Regional Science Association*, Vol. 10 (1962), 107-133; G. Rushton, R. G. Golledge,
and W.A.V. Clark, "Formulation and Test of a Normative Model
for the Spatial Allocation of Grocery Expenditures by a Dispersed Population," *Annals of the Association of American Geographers*, Vol. 57, No. 2 (1967), 389-400.

4.   References on this topic are summarized in
B.J.L. Berry and A. Pred, *Central Place Studies* (Philadelphia:
Regional Science Research Institute, 1965), 105-110 and in
G.Olsson, *Distance and Human Interaction* (Philadelphia: Re-
gional Science Research Institute, 1965), 8-10, 43-68.

5.   A comprehensive group of references is contained in
B.H. Stevens and C.A. Brackett, *Industrial Location* (Philadelphia: Regional Science Research Institute, 1967).

6.   Such assumptions are integral parts of the theories of location of both Christaller and Lösch, and they also are implicit in rent theory models and trade area models.   For lists of relevant references see Berry and Pred, *op. cit.* (1965), 97-109.

7.   J. Wolpert, "The Decision Process in a Spatial Context," *Annals of the Association of American Geographers,* Vol. 54 (1964), 537-58.

8.   P. Gould, "A Bibliography of Space-Searching Procedures," (unpublished manuscript, Department of Geography, Pennsylvania State University, 1966).

9.   R.G. Golledge and L.A. Brown, "Search, Learning, and the Market Decision Process," *Geografiska Annaler,* Vol. 49, No. 2 (1967), 116-124.

10.   D. Huff, "A Topological Model of Consumer Space Preferences," *Papers and Proceedings of the Regional Science Association,* Vol. 6 (1962), 157-173.

11.   D. Marble, "A Theoretical Exploration of Individual Travel Behavior," in W.L. Garrison and D.F. Marble (eds.), *Quantitative Geography* Northwestern University Studies in Geography No. 13, Evanston, Illinois (1967).

12.   A. Pred, *Behavior and Location* (Lund: Gleerup, 1967).

13.   G. Katona, *The Psychological Analysis of Economic Behavior* (New York: McGraw-Hill, 1951), 139.

14.   C.L. Hull, *A Behavior System* (New York: Wiley, 1964), 228.

15.   For a summary of these references see L. Brown, *Diffusion Dynamics: a Review and Revision of the Quantitative Theory of the Spatial Diffusion of Innovation* (Lund: Gleerup, 1968).

16.   Kurt Lewin, *Principles of Topological Psychology* Trans. by F. Heider and G.M. Heider (New York: MaGraw-Hill, 1936);   and *ibid, Field Theory in Social Science* (New York: Harper, 1951).

17.   K. Lewin, *A Dynamic Theory of Personality* Trans. by D.K. Adams and K.E. Zener (New York: McGraw-Hill, 1935), 88-91.

142

18.   W. Tobler, "Numerical Map Generalization," Discussion
Paper of the Michigan Inter-University Community of Mathematical Geographers, No. 8 (1966); A. Getis, "The Determination of
the Location of Retail Activities with the Use of Map Transformation," *Economic Geography*, Vol. 39 (1963), 14-22;
P. Gould, "On Mental Maps," Discussion Paper of the Michigan
Inter-University Community of Mathematical Geographers, No. 9
(1966).

19.   E.C. Tolman, *Purposive Behavior in Animals and Man*
(New York: Appleton-Century-Crofts, 1932); E.C. Tolman,
B.F. Ritchie and D. Kalish, "Studies in Spatial Learning II.
Place Learning versus Response Learning," *Journal of Experi-
mental Psychology*, Vol. 36 (1946), 221-229.

20.   E. Hilgard, *op. cit.* (1956).

21.   C.H. Hanzik and E.C. Tolman, "The Perception of Spa-
tial Relations by the Rat: A Type of Response Not Easily
Explained by Conditioning," *Journal of Comparative Psychology*,
Vol. 22 (1936), 287-318.

22.   E.R. Guthrie, *The Psychology of Learning* (New York:
Harper (Revised) 1952); *ibid.*, "Conditioning: a Theory of
Learning in Terms of Stimulus, Response, and Association,"
Ch. 1, in *The Psychology of Learning* National Social Studies
Education 41st Yearbook, Part II (1942), 17-60.

23.   E.R. Guthrie, "Association and the Law of Effect,"
Psychological Review, Vol. 47 (1940), 127-148.

24.   W.K. Estes et al., *Modern Learning Theory* (New York:
Appleton-Century-Crofts, 1954); *ibid.*, "Toward a Statistical
Theory of Learning," *Psychological Review*, Vol. 57 (1950),
94-107.

25.   C.L. Hull, *op. cit.*, (1964).

26.   C.L. Hull, *op. cit.*, (1964), 6.

27.   E.L. Thorndike, *The Fundamentals of Learning* (New
York: Teachers College, 1932).   Summaries of these laws are
given in E. Hilgard, *op. cit.* (1956), 27-30.

28.   C.L. Hull, *op. cit.* (1964), 253.

29.   R.S. Yuill, "A Simulation Study of Barrier Effects in
Spatial Diffusion Problems," Discussion Paper of the Michigan
Inter-University Community of Mathematical Geographers, No.5
(1965).

143

30. An extensive bibliography of relevant articles is given in L. Brown, *op. cit.* (1968).

31. See Hilgard, *op. cit.* (1956), 67-69.

32. R.M. Thrall, G.H. Coombs, and R.L. Davis (eds.), *Decision Processes* (New York: Wiley, 1954).

33. D. Huff, *op. cit.* (1962); D.J. Thompson, "Future Directions in Retail Area Research," *Economic Geography*, Vol. 42 (1966), 1-18; and J. Wolpert, *op. cit.* (1964).

34. S. Siegel et al., *Choice, Utility, and Strategy* (New York: McGraw-Hill, 1964).

35. C.L. Hull, *op. cit.* (1964); K.W. Spence (ed.), *Behavior Theory and Learning* (Englewood Cliffs, New Jersey: Prentice-Hall, 1960); C.J. Burke and W.K. Estes, "A Component Model for Stimulus Variables in Discrimination Learning," *Psychometrika*, Vol. 22 (1957), 133-145.

36. F. Restle, *Psychology of Judgement and Choice* (New York: Wiley, 1961); M. Levine, "Mediating Processes in Humans at the Outset of Discrimination Learning," *Psychological Review*, Vol. 70 (1963), 254-276.

37. S.H. Britt, The *Spenders* (New York: McGraw-Hill, 1960); L.C. Clark (ed.), *Consumer Behavior -- the Dynamics of Consumer Reactions* (New York: New York University Press (4th Printing), 1966); R. Ferber and H.G. Wales, *Motivation and Market Behavior* (Homewood, Illinois: R.D. Irwin, 1958); L.A. Fourt and J.W. Woodlock, "Early Prediction of Market Success for New Grocery Products," *Journal of Marketing*, Vol. 26 (October, 1960); R. Frank, "Brand Choice as a Probability Process," *Journal of Business*, Vol. 35 (January, 1962); A.A. Kuehn, "Consumer Brand Choice as a Learning Process," *Journal of Advertising Research*, Vol. 2 (December, 1962), 10-17.

38. B.J.L. Berry, H. Barnum, and R.J. Tennant, *op. cit.* (1962).

39. R.G. Golledge, G. Rushton, and W.A.V. Clark, *op. cit.* (1967).

40. This experiment (and some of the others reported in this paper) was carried out by R. Briggs and D. Demko, graduate students in the Department of Geography, Ohio State University, in a special seminar on learning models in April 1968.

41.   G. Rushton, "On the Scaling of Locational Preferences," *This volume*.

42.   R. Briggs, *The Scaling of Preferences for Spatial Locations: An Example Using Shopping Centers* (M.A. Thesis, Ohio State University: Department of Geography, 1969).

43.   Expositions of this view can be found in L. Postman, "One Trial Learning," in C.N. Cofer and B.S. Musgrave (eds.), *Verbal Behavior and Learning* (New York: McGraw-Hill, 1963), 295-333;   and B.J. Underwood and G. Keppel, "One Trial Learning?", *Journal of Verbal Learning and Verbal Behavior* 1   (1962), 1-13.

44.   This model is expounded at length in R.R. Bush and F. Mosteller, *Stochastic Models for Learning* (New York: Wiley, 1955).

45.   For example, see: M.F. Norman, "Incremental Learning on Random Trials," *Journal of Mathematical Psychology* 1 (1964), 336-350.

46.   R.R. Bush and F. Mosteller, *op. cit.* (1955).

47.   See R.G. Golledge, "Conceptualizing the Market Decision Process", *Journal of Regional Science* (Supplement), Vol. 7 (1967), 239-258.

48.   For references see Stevens and Brackett, *op. cit.* (1967).

49.   M. Shubik, *Strategy and Market Structure* (New York: Wiley, 1959);   and *ibid.*, *Game Theory and Related Approaches to Social Behavior* (New York: Wiley, 1964);   and H.O. Nourse, *Regional Economics* (New York: McGraw-Hill, 1968).

50.   R.D. Luce, "A Probabilistic Theory of Utility", *Econometrica*, Vol. 26 (1958), 193-224;   R.D. Luce, R.R. Bush, and E. Galanter (eds.), *Readings in Mathematical Psychology* Vol. 1 (New York: Wiley, 1963);   S. Siegel, "Theoretical Models of Choice and Strategy Behavior," *Psychometrika*, Vol. 24 (1959), 306-316;   H.A. Simon, "A Comparison of Game Theory and Learning Theory," *Psychometrika*, Vol. 21 (1956), 267-272.

51.   See D.L. Huff, *op. cit.* (1962);   J. Wolpert, *op. cit.* (1964).

52. R.D. Luce, *Individual Choice Behavior: a Theoretical Analysis* (New York: Wiley, 1959); W.S. Torgerson, *Theory and Methods of Scaling* (New York: Wiley, 1958); C.H. Coombs, *A Theory of Data* (New York: Wiley, 1964).

53. C.L. Hull, *op. cit.* (1964); K.W. Spence, *op. cit.* (1960); S. Siegel, *op. cit.* (1959); H.A. Simon, *op. cit.* (1956); R.C. Atkinson, *op. cit.* (1961); Patrick Suppes, "Behavioristic Foundations of Utility," *Econometrica*, Vol. 29 (1961), 186-202; J.B. Kruskal, "Non-metric Scaling: A Numerical Method," *Psychometrika*, Vol. 29 (June 1964), 115-129; R.N. Shepard, "Metric Structures in Ordinal Data," *Journal of Mathematical Psychology*, Vol. 3 (July 1966), 287-315.

54. Luce, *op. cit.* (1963).

55. This conclusion is based only on the experiments conducted at Ohio State University. Further experimentation may alter this conclusion.

56. W.K. Estes, *op. cit.* (1954); K.W. Spence, *op. cit.* (1960); R.C. Atkinson, *op. cit.* (1961); D.L. LaBerge and A. Smith, "Selective Sampling in Discrimination Learning," *Journal of Experimental Psychology*, Vol. 54 (1951), 423-430.

57. Luce, *op. cit.* (1963).

58. R.G. Golledge and L. Brown, *op. cit.* (1967); G.H. Haines, "A Theory of Market Behavior after Innovation," *Management Science*, Vol. 10 (1964), 634-655.

59. P. Gould, *op. cit.* (1966).

60. J. Wolpert, *op. cit.* (1964); P. Gould, "Wheat on Kilimanjaro: the Perception of Choice within Game and Learning Theory Frameworks," *General Systems Yearbook* (1967), 157-166.

61. See R.G. Golledge, "Conceptualizing the Market Decision Process," *Journal of Regional Science*, Vol. 7 (Supplement) (1967), 239-258.

62. Brown, *op. cit.* (1968).

Acknowledgement: Throughout the sections devoted to the presentation of learning models, continued recourse was made to the useful summaries contained in R.C. Atkinson, G.H. Bowers, and E.J. Crothers, *An Introduction to Mathematical Learning Theory* (New York: Wiley, 1956).

THE GENESIS OF ACQUAINTANCE FIELD SPATIAL STRUCTURES:

A CONCEPTUAL MODEL AND EMPIRICAL TESTS[1]

Kevin R. Cox

Ohio State University

Introduction

Contemporary geographical literature has recognized the importance of flows of information in decision-making behavior affecting a variety of spatial patternings. Particular emphasis has been placed upon interpersonal information flow within informal social contact networks or acquaintance circles.[2] The exact spatial characteristics of these interpersonal contact nets and the acquaintance circles which are manifestations of these nets, however, have received very little attention. It seems, therefore, that the investigation of such acquaintance networks is an important task which might well throw additional light upon processes underlying a wide variety of spatial patterns.

In addition, the spatial properties of acquaintance networks have an intrinsic geographical interest of their own. This is evidenced by the growing literature upon networks with a large number of contributions from fluvial geomorphology and transportation geography.[3] The study of acquaintance nets

146

might add a stimulus to network research as a result of iso-
morphisms hitherto unrealized.

Essentially acquaintance nets, as in the case of other
networks, indicate relationships between nodes. Such nets can
clearly be examined in a variety of ways. One might, for
example focus upon the dyadic relationships existing between
nodes, upon the acquaintance field of a given individual (node)
or upon clique structure. The focus in this paper is upon the
individual acquaintance field constituting those nodes receiving a choice from some originating node or source.

The primary intent of the paper is to establish guidelines towards the development of a plausible conceptual model
of acquaintance field formation in a spatial context. Such a
model should be consistent with current theory regarding spatial behavior. Also, it should be consistent in its predictions with real-world acquaintance field spatial structures.
The conceptual model and a test of the predicted spatial structures is set forth in the first section of this paper. A
secondary intent is to attempt an explanation -- consistent
with the conceptual model -- of the observed spatial structural
variation across individuals. This problem is treated in the
second section of the paper.

A. The Conceptual Model and Spatial Structures

Theory

The development of an acquaintance field in a spatial context can best be understood in terms of the individual relationships which, when aggregated, constitute the field. In
summary, two behavioral mechanisms are involved: first, the
acquisition of information in the form of perceptions of individuals; and second, the evaluation of the content of those
perceptions for the costs and rewards likely to be experienced
in a relationship. Preceding and then accompanying an

interpersonal attraction between two individuals is a flow of
information. Resulting from this transaction are mutual perceptions which form the basis for an evaluation of the rewards
and costs associated with or likely to be associated with a
continuation of the relationship.[4] Particularly important is
the transaction cost itself. Thus effort must be expended in
communication and this effort will be weighed against the
actual or anticipated rewards in deciding whether or not contacts with some other individual should be made or continued.
Clearly, to the extent that net rewards are perceived to be
high the interpersonal contact is likely to develop into an
acquaintanceship.[5]

The idea of expected or anticipated rewards or costs is an
important one, however, and calls for some clarification.
Direct experience of an individual and the accumulation of
knowledge about that individual allows evaluation of the relative costs and rewards likely to be associated with him. This
perception of an individual's ability to perform adequately in
the future, however, is contingent not only upon such direct
experience but also upon characteristics of the individual in-
ferred from the directly experienced. The knowledge that a
person comes from Mississippi or alternatively that he does
not come from one's neighborhood, allows one to infer a number
of other qualities about that individual.

The dynamic implications of the interpersonal contact for
possible acquaintanceship can be developed further in the same
way that Golledge and Brown have developed the implications of
early undirected search effort for ultimate stereotyped behavior in the case of the market decision process.[7] By way of
example, consider the case of a new resident: in the course of
searching for the high net rewards he desires from his interpersonal contact behavior a number of possible contacts will be
tried. From the collation of the net rewards obtained in such

sequential contacts, the individual will learn which contacts provide the greatest net reward: such contactees will tend to become the focus of the new resident's search efforts to the neglect of the less rewarding contacts. Hence, by a trial and error process in which information of either a directly experienced or inferential character is acquired and evaluated the new resident gradually maximizes the net rewards obtained from his contacts. Such equilibrium patterns of friendship, however, are clearly only apparent after the learning process has been completed. Actual maps of clique structure therefore, may contain not only equilibrium friendship patterns but also unstable friendship patterns associated with individuals at an earlier stage in the learning process.

The information transacted between two individuals, therefore, varies both quantitatively and qualitatively. In a quantitative sense *ego* acquires varying amounts of information about *alter* ranging from zero at one end of the scale, through varying degrees of uncertainty to, ideally, complete certainty. Qualitatively the information which is acquired has a content which can be evaluated in terms of its cost-reward implications.

From the geographer's viewpoint it is important to determine how the process of information acquisition and the content of that information which is to be evaluated can be related to location in space. Quantitatively the acquisition of information has a spatial dimension in terms of the relative locations of the source of information and the acquirer of the information respectively: relative centrality, intervening distance, for example, are factors which likely exercise a constraining effect upon such acquisition. Qualitatively, much of the personal information which one acquires is spatial in nature. An individual's locational center of gravity and related spatial behavior implies a great deal, both directly and inferentially about the rewards and costs likely to be experienced in a prolonged relationship. Hence, relative location can be seen as affecting not only the chance that one has information about an individual (i.e. information acquisition) but also as influencing the likely rewards or costs associated with that particular individual (i.e. information evaluation). The spatial correlates of information acquisition and of the information content which is to be evaluated will now be examined in greater detail.

In the process of acquiring information one can imagine two broad types of spatial situation:
1) Receipt of information about or from another individual is a function of distance-direction relationships. Relationships with intervening distance, frequently expressed probabilistically, are well represented in the geographical literature. Chance contacts over space apparently manifest a distinct localism partly as a result of the localization of many social activities which in turn derives from a desire to minimize energy expenditure.[8] Quite independently of distance minimization goals, however, an individual is likely -- for purely geometrical reasons -- to interact with a greater proportion of the population in his immediate neighborhood than with those further away.[9]

The impact of direction upon the probability of interpersonal contact has been less studied though there is sufficient evidence of an empirical nature[10] and also of a theoretical nature[11] to justify consideration here. Direction is considered in this section since the interaction effects between it and distance are regarded as particularly critical: more specifically we regard directional bias in an individual information field as being a result of or as being preserved by high movement costs associated with distance and therefore as stemming directly from the least effort postulate. Two points are relevant here: first, for an arbitrarily selected individual, trips will show a variety of lengths. For some individuals some spatial behaviors will be over particularly long distances due presumably to the high reward available at the destination and to the absence of intervening opportunities offering similar reward; shopping trips, marketing trips or the journey-to-work may all be considered in this light. The second point is that such long trips have two effects upon contact behavior and the acquisition of information about individuals: i) they impose a heavy opportunity cost such that movements in other directions over similarly long distances are less likely; ii) they increase the probability of contact between the trip maker and intervening locations relative to the probability of contact with locations which lie outside of that directional vector. Both i) and ii) can be effective in promoting a directional bias upon the information gathering of the individual making relatively lengthy trips.

2) Receipt of information about another individual is a function of belonging to the same acquaintance network. Operationally such acquaintance network bias in the formation of an acquaintanceship has been defined as either a sibling bias or as a grandparent bias. Such biases however must be defined in terms of the sociometric data upon which they are biased. Sociograms are N x N matrices with values S_{ij} equal to 1 or 0; an entry of unity indicates that j is chosen or is 'targeted' by i as a friend; zero entries indicate that j is not chosen as a friend. In a sibling-biased friendship or information flow the probability of a relationship between two individuals x and y is a function of the probability that they are both friends of, i.e. that they are both chosen by, a third party z, such that $S_{zx} = 1$ and $S_{zy} = 1$. In a grandparent-biased friendship or information flow the probability of a relationship between x and y is a function of the probability that x is friendly with a third party z who is also friendly

with y: i.e. $S_{xz} = 1$ and $S_{zy} = 1$. Thus whereas in the sibling-biased situation x and y are both friends of z, in the grand-parent-biased situation y is a friend of a friend z: this therefore describes a type of intervening opportunity situation in which z is an intervening opportunity or gate-keeper.

Qualitatively the information transacted in an interpersonal contact has certain spatial attributes which the receiver can perceive, and evaluate relative to his own purposes and also manipulate for the evaluatory inferences it can yield. The idea of perception is emphasized here. The locational attributes of the individuals involved in the relationship are important items of information in the evaluation of possible net reward levels associated with a relationship. Perceptions of these attributes are opposed to objective relative location are important not simply because of the distortion of a locational information by the individual receiver but also as a result of purposeful distortion by the sender of the information.[13]

1) The perception of relative location in terms of distance and direction has important effects upon the costs and rewards to be expected from a relationship. The least effort arguments with respect to distance should again be clear: friendships with individuals at a distance are costlier to maintain and such cost, whether measured in terms of time, opportunity cost, mean distance per contact or whatever, will become increasingly apparent as learning proceeds and cognition of distance relationships improves. A simple test of this hypothesis would be to examine the fortune of acquaintanceships following a spatial separation occurring at some point in time. As Thibaut and Kelley have pointed out, the well known dissolution of dating, "going steady" and engagement relationships with distance lend plausibility to this hypothesis.[14] Perceptions of direction, moreover, modify such an equation

between distance and effort since directional economy in the
making of contacts can permit multiple-contact trips with the
average movement effort expended per contact not much greater
than if all the contacts were distributed at short distances but
symmetrically around the locational center of gravity of the
contact-making individual. Especially common here is the maintenance and even growth of a large number of acquaintances in a
former area of residence: despite spatial separation between individual and acquaintances, the clustering of the acquaintances
permits multiple-contact movement. Such long distance multiple-
contact movement, however, will likely be at the expense of
developing acquaintanceships in other directional sectors.

Finally, there is the question of perceptions of locational
behavior. Where the movement patterns of two individuals in-
tersect frequently in both space and time, there is ample opportunity for the formation of acquaintanceships at relatively
lower movement cost than would be possible otherwise. In this
sense, schools, places-of-work, clubs, taverns, etc. become
an important element in the spatial process of acquaintance
formation.[15]

2)   Individuals not only have locations in a metric space,
however; they also have locations in a topological space and
mutual perceptions of these locations will color expectations
of reward and cost. *Alter's* location in a network of social
relations implies rewards for *ego* for a number of reasons.
Most obviously *alter* may be rewarding to *ego* since he performs
the function of an intervening opportunity by means of which
contact can be made with some third individual. As a corol-
lary statement *alter* may be rewarding due to his ability to
provide contacts with third, fourth, and nth parties for some
desired group activity.

These ideas, however, do not assume the attachment of any
values to the links in the social relations network: all links

are expressions of acquaintanceship and therefore express some
degree of liking. In actual fact links can be drawn between
all individuals and characterized in three ways: i) a relationship of positive interpersonal attraction, i.e. one of liking;
ii) a relationship of negative interpersonal attraction, i.e.
one of dislike; iii) a contact which has yet to be consummated.
Perceptions of locations can then be seen to extend to the
quality of the connections enjoyed by different participants
in a social relations network. In the light of the suggestions of cognitive balance theory[16] this type of information
is important for the recipient. Heider's theory of cognitive
balance, formalized in terms of signed digraphs by Cartwright
and Harary,[17] provides a mechanism for change in acquaintance
net configurations. In Heider's theory a triad is said to be
balanced if all relationships are positive or if two of the
relationships are negative. Thus if we take the point of view
of *ego* and if *ego* likes *alter*, *alter* likes *x* and *x* likes *ego*,
the situation is balanced. However, if *ego* likes *alter*, *alter*
dislikes *x* and *ego* likes *x*, the situation is unbalanced. *Ego*
is prompted by the resulting tension to change his orientation
towards *x* or towards *alter*. Heider's general theory in
behavioral terms is that persons tend to maintain similar
attitudes toward a person and toward things he is seen to
cause, possess or like.

Perceptions of network location in terms of the quality
of the connections, therefore, are important in the interpersonal contact and acquaintance formation process in that
they provide a core around which one's view of another person
is organized: i.e. the congruence or incongruence of persons
liked and disliked by the contacter and contactee becomes
highly critical information to which other information is
assimilated to form an organized, internally consistent view
of the other person's ability to perform adequately.

The process of acquaintanceship formation, therefore,
seems to be related to location in two ways: i) in the process
of acquiring information; the chance of acquiring information
about a person seems to be related to such locational attributes as intervening distance and acquaintance network affiliation; ii) in the evaluation of that information, perceptions
of relative location appear to be critical. We have also
developed the notion that relative location can be defined in
two ways: metric nearness as measured by distance and direction
relationships; and topological nearness as measured by connections in a system of interpersonal contacts.

The combination of these two subprocesses -- information
acquisition and locational evaluation -- with the two categories of locational constraint allow the definition of four
ideal-types of acquaintanceship formation in a spatial context: a) information acquisition constrained by distance-
direction relationships; evaluation in terms of distance-
direction; b) acquisition constrained by distance-direction;
evaluation in terms of connection; c) acquisition constrained
by connection; evaluation in terms of distance-direction;
d) acquisition constrained by connection; evaluation in terms
of connection.

Assuming that these processes are additive in their
effects,[18] one can suggest two axes of variation along which it
might be possible to array acquaintance fields in terms of
their spatial structures: (i) Distance-Biased Acquaintance
Fields: in the development of these fields either distance-
direction constraints or connection factors are important in
the acquisition of information (ideal-types a) and c) above).
Perceptions of relative location in metric terms, however, are
paramount in the estimation of anticipated rewards and costs.
Effort-minimization considerations seem especially important
here. In terms of spatial structure one would expect such

fields to be either short distance with negligible directional bias or long distance with strong directional bias. ii) Network-Biased Acquaintance Fields: in this case connection is the critical evaluatory criterion (ideal-types b) and d) above). Here, therefore, we are describing fields which vary in their connectivity; the more connected fields, moreover, should be those exhibiting acquaintanceships between individual and field members which are consistent with the hypotheses of sibling-biased and grandparent-biased information flows respectively. Clearly the criterion of acquaintance bias also entails a high proportion of acquaintanceships between individual and field members which are reciprocated. The empirical existence of such dimensions of variation will now be investigated with a factor analytical model based on sociometric data regarding choices of individuals as friends.

An Empirical Test

The data which form the basis for the empirical study are derived from sociometric data for a sample of 221 Swedish farmers belonging to an agricultural organization and living in the general neighborhood of Horby parish about forty kilometers northeast of Lund. The data were collected between 1964 and January 1965.[19] Every farmer was mailed a questionnaire asking him to check on a list of all other farmers in the neighborhood those with whom he enjoyed some sort of acquaintanceship. The questions were designed to elicit varying intensities of interpersonal attraction. The first question, for example, asked the farmer to mark the names of those with whom he probably spoke at least a few times during the year he became a member of the organization. The fourth question, on the other hand, asked him to take all those mentioned under the three preceding questions and detail those with whom he most frequently discussed or used to discuss farm problems.

In the analysis reported upon here, attention is confined
to relationships listed under question four, and to those cases
where the farmer listed five or fewer acquantances in toto over
all sociometric questions. This was suggested by the original
investigator after examining the statistical measures associated with the five or fewer farmers relative to statistical measures associated with friends chosen on question four. This
analysis concentrates upon all those who named four or more
friends. Of these, there were fifty-one.

For each acquaintance field involving the friendship
choices between an individual, his field members and among the
field members, the following measurements were made: i) The
mean distance separating the individual from his field members.
This variable was logarithmically transformed. ii) The directional bias of the field relative to the individual: this was
measured as the distance between the location of the sender or
source and the centroid of the acquaintance field. As this distance increases so does the spatial asymmetry of the acquaintance field relative to the individual. This variable also was
logarithmically transformed. iii) The connectivity of the
acquaintance field: connectivity has been defined as the total
number of acquaintanceships in the field as a proportion of all
N (N-1) possible relationships. This type of measure in which
the actual number of relationships is related to the total possible is analogous to some indices used in studies of transportation networks.[20] iv) Sibling bias: the proportion of the
individuals; acquaintanceships which are sibling-biased.
v) Grandparent bias: the proportion of the individual's acquaintanceships which are grandparent biased. vi) Reciprocity:
the proportion of the individual's acquaintanceships with his
field members which are reciprocated.

A factor analysis upon the resultant fifty-one by six matrix was performed and the factors rotated to a varimax position.

The resultant loadings for the factors with eigen-values greater than 1.0 are shown in the table below.

TABLE 1

Rotated Factor Matrix

	Factor One	Factor Two
Mean Distance	-0.1045	-0.9083
Directional Bias	-0.2865	-0.8842
Connectivity	0.8607	0.1775
Sibling Bias	0.8886	0.1665
Grandparent Bias	0.6716	0.1180
Reciprocity	0.6464	0.1691
Eigen value	3.02	1.18

Factor one, explaining just over 50% of the total variance emerges quite clearly as a network-biased dimension; factor two, explaining just short of 20% of the total variance is equally clearly the distance-biased dimension as predicted by the conceptual model.

B. Further Development of the Model

Thus far we have made various assumptions about behavioral mechanisms and deduced resultant acquaintance field spatial structures. The factor analytic model has confirmed the existence of these spatial structural types and is therefore consistent with theory but it is far from proving it. In particular individuals are seen to vary in the degree to which their acquaintance fields are either distance-biased or network-biased. The utility of our theory would be greatly enhanced if it could be extended to explain these variations. Such theoretical derivations could provide tests in the form of cross-sectional predictions. Such extension is the aim of the second section of this paper.

Theory

One possible explanation of spatial structural variation is that the field dimensions are related to each other temporally. It is quite possible, for example, that a more distance-biased network is characteristic of individuals at earlier stages in the process of developing an acquaintance field. Acquaintances acquired at this earlier stage could then provide the connections for the development of a network-biased field at a later stage. Consistent with this explanation are the following: i) the population of farmers is in the process of continual replacement either as a result of death and family succession or as a result of migration processes; hence it is logical to expect some individuals to be at an earlier stage in the acquaintance field development process than others. It is also plausible to assume that such replacement does not exhibit any spatial autocorrelation; hence one can regard farmers at later stages in acquaintance field development as forming an environment for those still in earlier stages. ii) For a new individual acquaintance field formation must commence with at least one distance-biased friendship since network-biased friendships are contingent upon already having an acquaintance.[21]

Any theory developed to explain such spatial structural variation, however, should be consistent not only with the conceptual model as developed thus far but also with theory regarding the spatial interaction between a set of actors and the location of interaction opportunities in space. Ullman discusses this problem in terms of three critical variables: a) complementarity; b) intervening opportunity; c) transferability.[22] In this particular problem the concept of complementarity is not relevant. The idea of intervening opportunities is relevant, however, as is the concept of

transferability. In this particular research context the
concept of intervening opportunities implies that more accessible opportunities for rewarding acquaintanceships will be substituted for the less accessible with "accessible" being defined in an ordinal sense. The concept of transferability
offers a cut-off point: the idea that interaction will only
take place with those rewarding opportunities if distance is
not so great as to impose prohibitive movement costs.

Our proposed extension of the conceptual model draws upon
all these considerations and places them in a geographically
variable environment in order for them to interact and produce
the observable field geometries. It is plausible to assume
that the values of the network-biased dimension vary across the
geographical area being considered; precisely why these variations occur is not known but we can present a hypothesis regarding their preservation over time. For example, as new
farmers replace older farmers in the area, their immediate
environment -- i.e. their neighborhood -- consisting of other
farmers offers a strong constraint upon the type of acquaintance field developed. In neighborhoods where acquaintance
fields are strongly network-biased -- a connection-rich environment -- an initial distance-biased friendship by the new
farmer is likely to result in a grandparent-biased information
flow situation allowing him to use the long-time resident as
an intervening opportunity in the development of his acquaintance field. On the other hand in neighborhoods where acquaint-
tance fields are less network-biased -- connection-poor environments -- the probability of gaining information about other possible acquaintances via an intervening opportunity is sharply
reduced. Given this type of situation in which accessibility
to a connection-rich environment is stressed, the distance-
biased acquaintance field appears as the normal state of the
system with deveiations from it resulting from acquaintanceships

made in a connection-rich environment. Spatial interaction
theory, with its emphasis upon intervening opportunity and
transferability is consistent with these notions.

An Empirical Test

In this research context we can measure intervening opportunity on a nearest-neighbor basis: the individual's score on
the network factor should, if our theory is valid, be a function of the network score for the nearest-neighbor, and to a
succeedingly lesser degree, a function of second-, third- and
nth-order nearest-neighbors. Transferability considerations
qualify this statement however. Employing distance (the only
available measure) as a measure of transferability, we would
anticipate an interaction effect between the nearest-neighbor
network score and intervening distance. The model to be
calibrated, therefore, is as follows:

$$Y_{i_c} = a + b_1 Y_{i1} + b_2 Y_{i2} + b_3 Y_{i1} D_{i1} + b_4 Y_{i2} D_{i2}$$

where Y_{i_c} = the predicted network-biased dimension
score for the ith observation

Y_{i1} = the network-biased dimension score for
the first-order nearest-neighbor of Y_i

Y_{i2} = the network-biased dimension score for
the second-order nearest-neighbor of Y_i

D_{i1} = the distance intervening between Y_i
and Y_{i1}

D_{i2} = the distance intervening between Y_i
and Y_{i2}

A stepwise multiple regression analysis provides the following
coefficient estimates:

$$Y_{i_c} = 0.421 Y_{i1} + 0.302 Y_{i2}$$

$$(3.59) \qquad (2.38)$$

These were the only coefficients significantly different from
zero at the ninetyfive percent significance level though between them the network-biased factor score for the first
nearest-neighbor and the network-biased factor score for the
second nearest-neighbor account for approximately 42% of the
original network score variance (R = .6447).   The relative
magnitudes of the two coefficients are consistent with a probabilistic interpretation of the intervening opportunities notion.   The nonsignificance of the other terms is a little
surprising though the geographical scales being considered
might not be such as to make transferability considerations of
critical behavioral significance.   Certainly the estimating
equation provides empirical evidence consistent with our theory
and therefore a point of departure for dynamic models aimed at
a spatio-temporal reconstruction of acquaintance fields.

C. Conclusions

Pairwise contacts are associated with a transaction of
information between the participant individuals.   This information is perceived by the individuals concerned and used to
evaluate the costs and rewards or the expectancies of cost and
reward likely to be associated with a renewal of the contact at
some future time.   Repeated contacts and eventual acquaintanceship result from a mutual evaluation of high net reward
as opposed to high net cost.

Location is seen as affecting this process in two ways.
Firstly, at the stage of information acquisition, the relative
locations of the individuals affect the chance that they will
know something about each other.   Secondly, at the stage of
information evaluation in which an individual takes his perceptions of the other and evaluates him in cost-reward terms,
perceptions of location can be seen as important components
in the individual's accounting scheme.

Location is viewed in two ways: firstly in metric terms, i.e. in distance-direction terms; and secondly in a topological sense as determined by the degree to which individuals have connections with each other. These considerations lead to the postulation of two axes of spatial structural variation along which acquaintance field spatial structures can be arrayed: a distance-biased dimension generated by an information acquisition process in which connection or distance and direction considerations are important and an information evaluation process in which perceptions of distance and direction relationships are critical: and a network-biased dimension in which the critical locational variables characterizing perceptions refer to such network characteristics as connectivity. A factor analysis of acquaintance field data for a sample of Swedish farmers suggested very strongly that indeed these two spatial structural dimensions did exist. Further analysis aimed at explaining the observed variation of these spatial structures across individuals led to the idea that the network-biased field is merely a special case of the distance-biased field. A strong neighborhood effect in the values of the network-biased acquaintance field score across individuals suggested that highly connected nets originate in distance-biased contact in a connection-rich environment.

The conceptual model is dynamic in nature and testing has involved the application of limited available data to an examination of cross-sectional predictions derived from the model. Clearly this is not wholly satisfactory from the point of view of relating process to structure. What is required is some model which will mirror the theorized dynamic processes and generate acquaintance field spatial structures. These spatial structures could then be compared with the real-world spatial structures identified in this paper and place the

conceptual model on a firmer theoretical foundation. An
appropriate model here would be spatial simulation not only in
the sense of mirroring a spatial process but also in the sense
of incorporating the feedback effects -- learning, the interaction effects between distance and direction and between network balance and acquaintance field extension -- which our
model suggests are important. Some future effort, therefore,
must be expended in this direction.

Possibly the most significant result emerging from this
research thus far therefore is the identification of distance-
biased and network-biased components in acquaintance field
spatial structure. This finding underlines a dichotomy which
Cavalli-Sforza applied to migration and which seems to have
general applicability to spatial behavior.[23] Cavalli-Sforza
discriminates between what he regards as the gravitational component and the diffusional component of migration. The gravitational component is characterized by a probability of selecting a new residential location which declines with distance
from the old residential location: he sees this as being
related to a localistic knowledge of area which in turn is
generated by least effort considerations: "...movements from
home, followed by return home due to work, search of food,
other type of duty or to entertainment. These will often be
highly repetitive: they may occasionally have only exploratory
purposes but they will, in any case, give to the individual a
knowledge of the area thus visited, and such knowledge may
eventually lead to marriage or change of residence."

The diffusional component on the other hand, explains
those residential site selections which bear little or no
relationship to distance from the previous site. Such a component appears to be related to the acquisition of information
which is not dependent upon considerations of minimization of

movement effort.   Rather information may be supplied by some
intermediary as in an informal communications network.

In current spatial behavior studies the significance of
the gravitational component is amply recognized and understood;
the diffusional component is less understood though apparently
it could be applied to a range of spatial behaviors other than
migration.   In acquaintance fields, for example, we have isolated a distance-biased or gravitational component and also a
network-biased component which appears to be independent of the
classic least effort considerations.   Likewise, in journey-
to-work studies it seems likely that there is not only a gravitational component to behavior but also a diffusional or network component resulting from, say, an information feedback
from former fellow employees.   Future work, therefore, should
be directed not only towards strengthening the validity of the
model insofar as it applies to acquaintance fields but also in
extending the conceptual generality of the model to cover a
wider range of spatial behaviors.

166

NOTES

1. The author would like to acknowledge the helpful comment and criticism of Dr Lawrence A. Brown, Department of Geography, Ohio State University upon an earlier draft of this paper.

2. See, for example: (i) Torsten Hägerstrand, *Innovation Diffusion as a Spatial Process* (Chicago: University of Chicago Press, 1967). (ii) Torsten Hägerstrand, "Migration and Area," in David Hannerberg, Torsten Hägerstrand and Bruno Odeving (eds.) *Migration in Sweden: A Symposium* (Lund: Lund Studies in Geography, Series B, Gleerup, 1957), 25-158. (iii) Lawrence A. Brown, *Diffusion Processes and Location* (Philadelphia: Regional Science Research Institute, 1968).

3. Peter Haggett, "Network Models in Geography," in R.J. Chorley and Peter Haggett (eds.), *Models in Geography* (London: Methuen, 1967).

4. This conceptualization in terms of the costs and rewards associated with an information flow between individuals relies heavily upon John W. Thibaut and Harold H. Kelley, *The Social Psychology of Groups* (New York: Wiley, 1967) especially Chapter 2.

5. Rainio has developed an empirical equation relating the emergence of an acquaintanceship to the intensity of interpersonal contacts: Kullervo Rainio, "A Study of Sociometric Group Structure: An Application of a Stochastic Theory of Social Interaction," in Joseph Berger, Morris Zelditch and Bo Anderson (eds.), *Sociological Theories in Progress* (Boston: Houghton Mifflin, 1965), 105.

6. Support for this notion comes from the psychology of interpersonal relations where it has been shown that during initial contacts various cues are presented; from these it is possible to make inferences about the ability of the other person to behave satisfactorily. See Thibaut and Kelley, *op. cit.*, 74.

7. Reginald G. Golledge and Lawrence A. Brown, "Search, Learning and the Market Decision Process," *Geografiska Annaler,* Vol. 49, Ser. B, No. 2, (1967).

8. See, for example: Richard L. Morrill and Forrest R. Pitts, "Marriage, Migration and the Mean Information Field: A Study in Uniqueness and Generality," *Annals of the Association of American Geographers,* Vol. 57, No. 2 (1967), 405-406.

9. James S. Coleman, *Introduction to Mathematical Sociology* (New York: Free Press of Glencoe, 1964), Chapter 15.

10. For example, Duane F. Marble and John D. Nystuen "An Approach to the Direct Measurement of Community Mean Information Fields," *Papers and Proceedings, Regional Science Association,* Vol. 11 (1963), 99-109.

11. The outstanding instance here is Torsten Hägerstrand (1957), *op. cit.*

12. See T.J. Fararo and Morris H. Sunshine, *A Study of a Biased Friendship Net* (Syracuse: Youth Development Center, Syracuse University, 1964).

13. Beyond a certain intervening distance, for example, it is perfectly possible for a sender to exaggerate his locational isolation if he wishes to terminate the embryonic relationship; likewise, encouragement can be provided by an overly optimistic assessment of locational convenience.

14. Thibaut and Kelley, *op. cit., 41.*

15. Katz and Hill have discussed the role of such places as sites for initial interpersonal contacts in marital selection and have called them *organization points.* Available data indicates a high correlation between organizational point propinquity and residential propinquity suggesting that measures of relative location in distance and direction terms provide an adequate surrogate for spatial behavior and the probability of contact at some organization point. This adequacy, however, is likely to be upset with increasing spatial concentration of organizational points. See Alvin M. Katz and Reuben Hill, "Residential Propinquity and Marital Selection: A Review of Theory, Method and Fact," in Jean Sutter (ed.) *Les Déplacements Humains* (Entretiens de Monaco en Sciences Humaines, May, 1962).

16 Fritz Heider, *The Psychology of Interpersonal Relations* (New York: Wiley, 1958).

168

17. D. Cartwright and F. Harary, "Structural Balance: A Generalization of Heider's Theory," *Psychological Review*, Vol. 63 (1956), 277-293.

18. We have no justification to think otherwise at present though this might be a legitimate area for future theory building.

19. The survey was organized and executed by Dr Lawrence A. Brown of the Department of Geography, Ohio State University. His willingness to make the data available to me is gratefully acknowledged.

20. An example is the gamma index developed for the planar graph isomorphic to (e.g.) a railroad network. See, for example, William L. Garrison and Duane F. Marble, *A Prolegomenon to the Forecasting of Transportation Development* (Evanston, Illinois: The Transportation Center, Northwestern University, 1965), 51.

21. It might be added that the fact that the distance-biased dimension explains less than half of the total field variation which is explained by the network-biased dimension suggests that for many the distance-biased field might be a relatively transient phenomenon.

22. Edward L. Ullman, "The Role of Transportation and the Bases of Interaction," in William L. Thomas (ed.) *Man's Role in Changing the Face of the Earth* (Chicago: University of Chicago Press, 1956).

23. L. Cavalli-Sforza, "The Distribution of Migration Distances: Models and Applications to Genetics," in Jean Sutter (ed.), *Les Déplacements Humains* (Entretiens de Monaco en Sciences Humaines, May, 1962).

ON THE IMPLEMENTATION OF PLACE UTILITY AND RELATED CONCEPTS:

THE INTRA-URBAN MIGRATION CASE[1]

Lawrence A. Brown

Ohio State University

and

David B. Longbrake

University of Iowa

Although place utility has been posited as a major component of behavioral conceptualizations of intra-urban migration processes, specification of place utility in empirical
and operational terms has not generally been accomplished.[2]
One task of a series of complementary intra-urban migration
research efforts in which the authors are involved has been to
examine approaches to operationalizing the concept of place
utility.[3] The design of those efforts is tailored to fit
within the ultimate goal of developing general models of residential site selection which incorporate well-grounded behavioral principles relevant to intra-urban migration processes.
As a practical matter, although a household level of aggregation would be desirable, it is recognized that an ecological
approach which focusses upon areal units within the city must
be taken, given the restrictions imposed by available (or non-
available) data.   Thus, such models must employ variables

acting as surrogates for behavior at a macro-scale, and be aggregative in nature.[4]

Within this framework two approaches have been examined. Initially, we attempted to construct surrogate utility functions using ecological data describing socioeconomic characteristics of areal units within a city. These functions were calibrated on the basis of actual migration flows in Cedar Rapids, Iowa for the year 1966 to 1967.[5]

The second approach is to construct operational models of migration which have an identifiable place utility component. The description and investigation of one such model constitutes the focus of this paper. First, however, we present a summary of a basic conceptual framework for migration decision processes which has been set forth elsewhere by Brown and Moore.[6] This framework provides the rationale for the model presented here.

Brown and Moore view the intra-urban migration decision in two phases: (1) The decision to seek a new residence; and (2) The decision of where to relocate. The latter involves two simultaneous endeavors -- the search for available vacancies and the evaluation of each vacancy encountered.

The first decision is seen as one of several responses of the household to the impact of stress. In terms of the present discussion, stress would result from dissatisfaction with the household's experienced place utility. In this context, the process of intra-urban migration may be seen as one of adjustment whereby one residence is substituted for another in order to better satisfy the needs and desires of the household, i.e., in order to increase its experienced place utility.

The outcome of the second decision, the choice of a new residence, depends upon the comparative place utility associated with each vacancy, and the migrant household will naturally choose to maximize its gain in experienced place utility.[7]

However, several vacancies may be similar in terms of their subjectively judged place utility. Furthermore, the opportunity of evaluating a particular vacancy by a particular migrant is completely dependent on its being found by that migrant. The selected new residence, in fact, must be recognized to be but one of several possible outcomes, each highly dependent upon the migrant household's search procedures.

The prime constraint on a migrant household's search pattern is provided by its awareness space, which may be defined as that set of locations within the urban area about which the migrant possesses some knowledge. Since, with respect to the individual migrant household, the urban area is also characterized by a variable surface describing the amount of information possessed by the household for each location, some locations within the awareness space are better known than others, and only those locations for which information possessed falls below some threshold value (possibly zero) will lie outside the awareness space.

Information relevant to the awareness is derived from two main sources: (1) The migrant household's activity space which is defined as that set of urban locations with which direct contact has occurred as the result of past activities; and (2) Its indirect contact space which is defined as that set of urban locations for which the migrant possesses information as the result of indirect contact through channels such as acquaintances or the mass media.

The actual search for vacancies is undertaken within the framework of the household's search space. The search space is contained within its awareness space and comprises those locations within the urban area which the migrant household perceives as being likely to satisfy its aspirations with regard to a new residence. Thus, Brown and Moore explicitly recognize that, on the basis of criteria relevant to the utility of a residential site (such as its accessibility to urban amenities or its social environment), certain locations within the urban area will be eliminated from consideration before search begins.

Time imposes an additional factor in the decision to seek a new residence. On the one hand, the passage of time may result in an increase in the volume of the awareness space and a change in the volume of the search space, either increasing or decreasing it depending upon the household's search experiences. On the other hand, as the time remaining to find a suitable new residence decreases, the household is subjected to increasing stress with the result that the ability to make accurate evaluations of encountered vacancies becomes severely curtailed. Thus, the relationship between awareness space, search space, and the associated search behavior is seen to be time dependent.

The remainder of the paper is divided into three segments. The first deals with those aspects of our operational model which relate to the generation of migrant households and vacancies within each areal unit in the city. In terms of the conceptual framework presented above this aspect of the model pertains to the decision to seek a new residence. The second segment of the ensuing discussion deals with those aspects of the model which relate to the allocation or distribution of migrant households among available vacancies. In terms of the conceptual framework presented above, this aspect of the model pertains to the decision of where to relocate. The third and final section below presents a commentary on the operational model and its relationship to the conceptual framework.

A. An Operational Format for Examining Place Utility

1. Generation of Migrant Households and Vacancies

Phase 1.1 of the model consists of estimating the number of households per areal unit or origin-destination (O-D) zone which will change their residence in time t and the number of vacancies per O-D zone which become available in time t.[8] The former estimate may be based upon empirical observations of the relationship between number or proportion of migrant households generated by an O-D zone and that zone's environmental and household characteristics.[9] Thus, letting $m_i(t)$ represent the number of out migrants from zone i during time t

$$m_i(t) = f(E_i(t), H_i(t)) \qquad (1)$$

where E and H represent variables or factors describing the environmental and household characteristics of O-D zone i during time t. Specific environmental and household characteristics which would be considered in estimating (1) are discussed by Brown and Longbrake, Rossi, and Simmons; functions of the type to satisfy equation (1) have been estimated for Cedar Rapids Iowa, by Brown and Longbrake.[10]

Estimation of the number of vacancies per O-D zone which are available by the end of time t, should be based upon the number of migrant households generated by each O-D zone, since each migrant household leaves a vacancy behind, and upon the number of newly built homes made ready for the market during time t. In addition, assuming that all vacancies in time t-1 are not filled immediately, there will be an inventory of vacancies for time t. Thus,

$$v_j(t) = v_j(t-1) - \sum_i x_{ij}(t-1) + m_j(t) + b_j(t) \qquad (2)$$

where $v_j(t)$ is the number of vacancies in zone j during time t,

X_{ij} (t-1) is the number of household migrations from zone i to
zone j during time t-1 (where i=j is possible), m_j(t) is as
defined above, and b_j(t) is the number of newly built dwelling
units appearing completed for occupancy in zone j during time
t.[11,12]

Phase 1.2 of the model consists of estimating the socioeconomic characteristics of migrant households and vacancies
for each zone. Solely on a zone basis, which would assume
that all migrants and vacancies in the zone are socioeconomic-
ally homogeneous, this must be accomplished by a factorial
ecology type of approach.[13] However, it is desirable to go a
step further and specify a distribution of migrant household
types and vacancy types for each type of zone. Thus, we have

$$m_{ik}(t,g) = m_i(t,g)\ P_{kg}(M) \tag{3}$$

and

$$v_{jk}(t,g) = m_i(t,g)\ P_{kg}(V) \tag{4}$$

where m_{ik}(t,g) indicates the number of migrant households of
socioeconomic category k generated in time t by zone i which
is of socioeconomic type g, P_{gk}(M) indicates the probability
that a migrant household is of socioeconomic type k if it is
found in zone type g, v_{jk}(t,g) indicates the number of vacancies of socioeconomic type k generated in time t by zone j
which is of socioeconomic type g, and P_{kg}(V) indicates the
probability that a vacancy is of socioeconomic type k if it
is found in zone type g.

As an illustration consider the socioeconomic classifi-
cations for Cedar Rapids 0-D zones which were identified in
the earlier study by Brown and Longbrake. Table 1 presents
descriptions of each of their five-zone types: (I) Middle
Life Cycle, Middle Class Family Households, (II) Late Life
Cycle, Upper Middle Class Households, (III) Lower Economic

Status, Sound Rented Two-Family Dwelling Units, (IV) Lower
Economic Status, Unsound Rented Multi-Family Dwelling Units,
(V) Downtown Commercial Dwelling Units.

The spatial distribution of these zone types for Cedar
Rapids, Iowa is approximately in a concentric circle pattern
with sectorial bias. By applying the criteria of these classi-
fications to selected responses to a questionnaire distributed among recent intra-urban migrants in Cedar Rapids in the
spring of 1967, estimates are also available for the distribution of migrant household types for each zone type (Table 2),
which provides an estimate for $P_{kg}(M)$, and for the distribution of vacancy types for each zone type (Table 3) which
provides an estimate for $P_{kg}(V)$.

2. Allocation of Migrant Households

A critical problem in specifying a mathematical allocation system for migrant households is to avoid assigning more
than one household per vacancy. Although this problem could
be handled in a Monte Carlo framework, linear programming initially provides a more interesting approach, given the simplified modeling system which data availability necessitates.
Furthermore, a linear programming formulation seems reasonable
in that migrant households evidently operate to maximize some
sort of place utility function while minimizing search effort,
and this takes place in a highly competitive and fragmented
market situation.[14, 15]

An important input to the linear programming model is the
cost of relocating a household from a residence at site i to
another at site j, symbolized as c_{ij}. In accord with the
conceptualization put forth by Brown and Moore, summarized
briefly above, c_{ij} is seen to be a function of the migrant
household's awareness space and aspiration set. The former
reflects the effort in a household at i searching out and

identifying a vacancy at j. The latter provides an indication
of the return or satisfaction accruing to a household from i
which takes a vacancy at j.

To operationalize the awareness space initially, a distance decay type function may be used to indicate the intended
migrant household's information levels for locations at varying distances from his present residence. As an example,
Marble and Nystuen present a function based upon 1949 urban
travel behavior in Cedar Rapids, Iowa, and Morrill and Pitts
present several other examples of a similar sort.[16] However,
a set of functions more relevant to the present research
effort can be derived from data collected in Cedar Rapids by
questionnaire methods in the spring of 1967 which disclose the
locations of acquaintances for approximately 200 households
that changed residence sites not more than one month before
the interview. This is currently being analyzed to calibrate
four separate sets of distance decay functions: one general
function employing all data, which is comparable to those
derived by Marble and Nystuen for Cedar Rapids; a set of
functions, each differentiated on the basis of the respondent's
location vis-a-vis the CBD; a set of functions, each differentiated on the basis of characteristics of the respondent household's environment, indicated by the O-D zone typology presented in Table 1; and a set of functions, each differentiated
on the basis of socioeconomic characteristics of the household
itself, again employing the typology presented in Table 1.
It appears that employing a "specialized" awareness space
function which reflects differing characteristics of households
is likely to yield a more satisfactory analysis than employing
a single "general" function, although firm conclusions must
await our tests. In any case, as a knowledge of urban awareness spaces increases, we expect that researchers will formulate more complex functions which take account of the

influence of areal characteristics other than its distance
from the migrant. Clearly, these would not be purely distance
decay in form, although they would probably contain a strong
distance decay component. Currently, however, the basis for
calibrating such a function does not exist.

For purposes of specifying the awareness space concept
within the allocation model, let a_{ijk} equal the probability
that a household of socioeconomic characteristics k located
in zone i has a-priori knowledge of vacancies and other characteristics of zone j. Then $1-a_{ijk} = a^*_{ijk}$ provides a measure
of the relative effort involved in finding a vacancy at j.
Thus, if j is totally outside i's awareness space, $a_{ijk} = 0$
and $a^*_{ijk} = 1$; if j is within k's awareness space,
$0 \leq a^*_{ijk} < 1$. If a_{ijk} is calibrated by a distance decay function,
as suggested above, a^*_{ijk} will increase monotonically to 1 as
the distance between i and j increases. If distance decay
functions differentiated on the basis of household socioeconomic characteristics are available, such as in the Cedar
Rapids example, the index k would signify which function should
be applied.

As noted, a second basic element in the relocation process is the intended migrant household's aspirations with respect to a dwelling and neighborhood. Several criteria
which might enter into defining an aspiration set were discussed in Rossi, Simmons, and in the earlier work of Brown and
Longbrake. These are reflected in the characteristics of
the socioeconomic categories described in Table I. The aspiration set is seen to function both in terms of (1) identifying locations within the awareness space which are most
likely to provide acceptable vacancies, and (2) deciding upon
one vacancy from those which the migrant household knows to
be available.

For purposes of the linear programming model, the aspiration set is represented by a variable d_{ijk} which provides a relative measure of the distance between place i and place j for a household of socioeconomic type k in terms of the expected aspirations of a k household in zone i, and the environmental or neighborhood characteristics of zone j. To estimate d_{ijk} the researcher could employ a function such as (1), using d_{ijk} as the dependent variable and variables to characterize E_{jk} and H_{ik} as independent variables.

As illustration of another approach again consider the Cedar Rapids study by Brown and Longbrake. A sample of approximately 1500 movers was disaggregated into a 5 x 5 matrix so as to indicate the number of movers leaving a zone of one socioeconomic type and going to a zone of the same or another socioeconomic type (Table 4-A), where the five zone types are those described in Table 1.[17] Following the suggestion of Beshers and Laumann, the movement matrix was then transformed into a Markov chain transition matrix (Table 4-B). which was subsequently transformed into a mean first passage time matrix (Table 4-C).[18] The elements of the latter can be seen as a measure of the relative distance between a zone of one socioeconomic type and a zone of another (or the same) socioeconomic type. This measure is based upon actual choice behavior by migrant households and takes into account the general socioeconomic characteristics of both the origin and destination place. It also has a number of other characteristics which are relevant to its use as a distance measure of the type required; these are discussed in detail by Brown and Horton. Thus, it appears that measures such as those presented in Table 4-B are suitable estimates for d_{ijk}, assuming that the migrant household's socioeconomic type has no effect upon household aspirations.[19] Comparable measures which do reflect the distribution of both vacancy and household

socioeconomic types for a given type of zone are presented in
Table 5.[20]

Other elements employed for the primal of the linear programming allocation model are $x_{ijk}(t)$, $m_{ik}(t)$ and $v_{jk}(t)$, all
of which were defined in the discussion of Phase I of the
model. Combining these into a linear programming format, the
primal may be stated as

$$M = \sum_i \sum_j \sum_k x_{ijk}(t) \, d_{ijk} \, a^*_{ijk}$$

subject to

$$\sum_j x_{ijk}(t) = m_{ik}(t) \qquad (5)$$

$$\sum_i x_{ijk}(t) = v_{jk}(t)$$

$$x \geq 0$$

which is immediately recognizable as the classical transporta-
tion problem under the assumption that $\sum_i \sum_k m_{ik}(t) = \sum_j \sum_k v_{jk}(t)$.[21]
Here, however, we minimize the product of migration distance
(in terms of household aspirations) and search effort subject
to the restrictions that (1) the number of households of type
k leaving zone i equal the number of households of type k
designated as out-migrants from zone i for time t and (2) the
number of households of type k assigned to zone j equal the
number of type k vacancies in zone j for time t. The criterion function and constraints of the linear programming primal
may be seen to take account of the establishment of a set of
aspirations, the seeking out of vancancies, the application of
the aspriation set to characteristics of those vacancies, and,
finally the selection of one. Incidentally, the designation
of c_{ij} as the product of migration distance (d_{ijk}) and search
effort (a^*_{ijk}) is somewhat arbitrary, and in light of the

interpretation of shadow prices given below, it might be more reasonable to define c_{ij} as $d_{ijk} - a_{ijk}$. Apparently, such a change would affect neither the form of the model nor the interpretive discussion below.

If we wish to consider the alternative that no suitable vacancy is found, adjustment of the model to include "slack" states is possible. Then, following through on the households which are allocated to slack states in time t would be interesting, and this is made possible by the recursive nature of the model, i.e., the model can be applied to successive time intervals. A stimulating idea made possible by this format and suggested by the work of Day and Schlager is to add one or more constraints so as to limit the behavior in one time interval according to that of the previous interval.[22] However, this exercise will be reserved for a later effort.

Recalling the discussion of the derivation of d_{ijk} and a^*_{ijk}, it is interesting to note that minimizing migration distance and search effort in (5) can be interpreted as the equivalent of maximizing the household's gain from migration as measured by the increase in realized aspirations tempered by the search effort involved in acquiring a suitable vacancy. In terms of the discussion above, this may be seen as maximizing the increase in experienced place utility. Following Stevens' suggestions, this point becomes clearer by considering one form of the dual of (5)

$$\text{Max} = \sum_j \sum_k v_{jk}(t) \, w_{jk} - \sum_i \sum_k m_{ik}(t) \, u_{ik}$$

subject to $\qquad\qquad\qquad\qquad\qquad\qquad\qquad\qquad$ (6)

$$w_{jk} - u_{ik} \le d_{ijk} a^*_{ijk}$$

where w_{jk} and u_{ik} represent the shadow prices associated with type k vacancies at zone j and intended migrant households of type k at zone i, respectively.[23] In the traditional

employment of the transportation problem, shadow prices are
interpreted as an abstract measure reflecting the value to the
system of having goods located at places i and j (or ik and
jk in the context of the present problem).   Thus, the amount
w_j - u_i represents the increase in value that results from
moving a unit of goods from place i to place j.   In terms of
our problem, as indicated in the constraint of (6), that increase is measured in terms of the gain in realized aspiration
level and the search effort (cost) involved in realizing that
gain, and w_{jk} - u_{ik} represents the increase in realized place
utility that results from moving a household of type k from
location i to a type k vacancy in location j.   To view this
in terms of the total system, we may rewrite the criterion
function of (6) as

$$\text{Max} = \sum_i \sum_j \sum_k x_{ijk}(t) \ (w_{jk} - u_{ik}) \qquad (7)$$

by substituting $\sum_j x_{ijk}(t)$ for $m_{ik}(t)$ and $\sum_i x_{ijk}(t)$ for $v_{jk}(t)$.
It then becomes apparent that the pattern of allocation of
households for the whole system (as represented by the set of
x_{ijk}'s greater than zero) is such as to maximize the increase
in system-wide place utility.

To take in the realities of the housing market, empirical
application of a model such as this should employ relatively
small time intervals, say a month, in a recursive format.
In such an exercise it should be profitable to examine both
the spatial and socioeconomic regularities manifest in the
normative migration patterns generated by the model, the deviations of actual migration patterns from normative ones, and
the utility measures w_{jk} and u_{ik}.   The latter task might also
include examination of regularities in the differences between
w_{jk} and u_{ik} for all combinations of origin-destination zones.
Clearly, such empirical work would lend insight into place

utilities, its aggregate measurement, and its role in the
migration process. However, it is also important to further
interpret location rent concepts, such as those developed by
Stevens, in the context of a migration format.[24] These would
provide further bases for establishing the relationship between
migration processes conceptualized in behavioral terms and
other processes of movement more often described in pure economic terms.[25]

B. Commentary

A basic task underlying the work reported here, as well
as the work reported earlier by Brown and Longbrake, is to
provide a format for operationalizing verbal behavioral concepts.[26] It is our contention that such operationalization
is ultimately necessary if the behavioral models of geographers are to be effectively implemented. In our particular
efforts we choose an aggregative ecological type framework
within which to operate, as a result of our interest in macro-
scale planning type models, but this emphasis is not necessary.

Whatever the emphasis, however, the task of operationalization involves moving from a conceptual model to a simplified
model, to borrow Chorley's terminology, and this necessarily
involves a loss of detail and richness.[27] The operational
format reported here is no exception to this rule. However,
we consider the operational and theoretical implications of
the model to be satisfactory, both from the point of view
that they represent an initial operational approximation of an
intra-urban migration system and that they provide insight into
the place utility concept, as well as our shortcomings in
knowledge on place utility and its operational role in intra-
urban migration processes. Such a shortcoming is particularly
obvious with respect to Phase I of the model, that dealing with
the decision to seek a new residence. Relatively speaking,

this phase is largely descriptive, rather than explanatory, and thus offers few insights into the workings of place utility with respect to households that are not yet declared migrants, as well as the conditions under which adjustment in situ is chosen as an alternative instead of seeking a new residence.[28] Clearly, this phase should be singled out for special attention in future research efforts.

TABLE 1

Socioeconomic Typology of O-D Zones: Cedar Rapids, Iowa

I.   Middle Life Cycle, Middle Class Family Households.

Mixed occupation middle income families ranging in income from $6,000 to $10,000.

Large, relatively young families with age groups 0 to 19 and 40 to 49.

High proportion of single family homes most of which are owner-occupied with very few vacancies and some high value and high rents.

High auto to population ratio, with car as the principle means of transportation to work.

Extensive tracts of underdeveloped land, especially in peripheral zones.

II.   Late Life Cycle, Upper Middle Class Households.

White collar high income families ranging in income from $10,000 to $25,000.

These zones are almost entirely residential with homes of good quality and high average value.

Families are of moderate size, but have a more advanced age component (a possible college age element of 20 to 29 and an associated paternal age group of 50 to 60+).

III.   Lower Economic Status, Sound Rented Two Family Dwelling Units.

Some large families with high densities per occupied unit and lower incomes from $3,000 to $6,000.

A sizeable percentage cf the housing units are two family structures resulting in higher population densities per acre.

A strong 20 to 29 age factor may suggest the presence of a newly formed family component.

TABLE 1 (contd.)

III.   (contd.)

These zones also contain an older age sector from
50 to 60+.

IV.   Lower Economic Status, Unsound Rented Multi-Family
       Dwelling Units.

These are the older deteriorating residential
sections of the city which have already sustained
considerable commercial land use invasion.

Most residential structures are rented and multi-
family occupied with high rates of vacancy and
dilapidation.   The dwelling units per acre ratio
is also high.

A significant number of incomes are less than
$3,000, and jobs include a high percentage of sales
and personal service people.   Many of these people
take the bus to work or make use of some means of
transportation other than car.

Unemployment is higher here than in any other part
of the city, and these zones also contain more than
a proportional share of the retired population.

V.   Downtown Commercial.

186

TABLE 2

Estimated Distribution of Migrant Household Types for Each Zone Type

Zone Types	Household Types			
	I	II	III	IV
I	.820	.049	.082	.049
II	.225	.425	.250	.100
III	.378	.054	.460	.108
IV	.235	.059	.275	.431

Sample size for each zone: type I = 61, type II = 40, type III = 37, and type IV = 51.

TABLE 3

Estimated Distribution of Vacancy Types for Each Zóne Type

Zone Types	Vacancy Types			
	I	II	III	IV
I	.723	.060	.193	.024
II	.361	.472	.111	.056
III	.273	.091	.545	.091
IV	.081	.000	.216	.703

Sample size for each zone: type I = 83, type II = 36, type III = 33, amd type IV = 37.

TABLE 4

MIGRATION DISTANCE ANALYSIS

A. MIGRATION MOVEMENTS BY NEIGHBORHOOD TYPE

		I	II	III	IV	V
from	I	240.	65.	11.	23.	1.
	II	118.	72.	8.	29.	0.
	III	44.	18.	24.	25.	0.
	IV	68.	37.	18.	73.	2.
	V	5.	6.	3.	9.	1.

B. TRANSITION PROBABILITIES BY NEIGHBORHOOD TYPE

		I	II	III	IV	V
from	I	.71	.19	.03	.07	.00
	II	.52	.31	.04	.13	.00
	III	.40	.16	.22	.22	.00
	IV	.34	.19	.09	.37	.01
	V	.21	.25	.13	.37	.04

C. MEAN FIRST PASSAGE TIMES BY NEIGHBORHOOD TYPE

		I	II	III	IV	V
from	I	1.66	5.28	24.92	11.39	326.35
	II	2.08	4.62	24.71	10.61	327.23
	III	2.45	5.47	20.02	9.23	326.86
	IV	2.60	5.32	22.80	7.81	323.05
	V	2.92	4.99	21.83	7.49	312.80

EQUILIBRIUM DISTRIBUTION

I	II	III	IV	V
.60	.22	.05	.13	.00

TABLE 5

MIGRATION DISTANCE ANALYSIS CONSIDERING BOTH
NEIGHBORHOOD TYPE AND HOUSEHOLD OR VACANCY TYPE

A. TRANSITION PROBABILITIES

Neighborhood Type	Household Type	I				II				III				IV			
		I	II	III	IV	I	II	III	IV	I	II	III	IV	I	II	III	IV
from I	I	.54	.04	.14	.02	.05	.07	.02	.01	.03	.01	.05	.01				.01
	II					.24	.31	.07	.04	.09	.03	.19	.03	.02		.04	.14
	III					.07	.09	.02	.01	.16	.06	.33	.06	.08		.22	.70
	IV																
II	I	.56	.05	.15	.02	.27	.36	.08	.04	.03	.01	.06	.01	.01		.02	.08
	II	.09	.01	.02						.03	.01	.07	.01		.01		
	III	.35	.03	.10	.01					.06	.02	.11	.02	.02		.07	.21
	IV									.07	.02	.14	.02	.06		.16	.53
III	I	.51	.04	.14	.02	.08	.10	.02	.01	.02	.01	.04	.01				
	II	.36	.03	.10	.01	.18	.23	.06	.03	.13	.04	.26	.04	.02		.05	.17
	III	.17	.01	.05	.01	.02	.02		.01	.07	.02	.14	.02	.04		.11	.35
	IV					.17	.24	.06	.03								
IV	I	.36	.03	.10	.01	.09	.12	.03	.01	.07	.02	.14	.02	.07		.18	.57
	II	.47	.04	.13	.02	.12	.16	.04	.02	.09	.03	.20	.03				
	III	.36	.03	.10	.01	.05	.07	.02	.01	.02	.01	.05	.01				
	IV	.06	.01	.02													

190

TABLE 5 (Continued)

B. MEAN FIRST PASSAGE TIMES

Neighborhood Type	Household Type	I				II				III				IV			
		I	II	III	IV	I	II	III	IV	I	II	III	IV	I	II	III	IV
from I	I	3.38	38.95	10.27	96.42	13.97	15.56	47.13	92.92	18.35	54.91	11.22	54.67	59.81	6858.11	22.42	14.34
	II	5.51	41.08	12.41	98.54	10.56	12.16	43.80	89.58	17.31	53.87	10.20	53.63	59.63	6857.89	22.18	14.12
	III	5.74	41.32	12.64	98.75	13.96	15.57	47.18	92.94	15.75	52.30	8.64	52.05	57.66	6855.89	20.19	12.12
	IV	6.44	42.03	13.34	99.36	16.48	18.08	49.69	95.43	19.18	55.77	12.07	55.51	50.72	6849.01	13.30	5.25
II	I	3.35	38.90	10.24	96.39	15.02	16.62	48.21	93.98	18.25	54.82	11.13	54.57	59.02	6857.32	21.60	13.55
	II	5.14	40.71	12.04	98.16	9.87	11.46	43.11	88.86	18.54	55.11	11.44	54.86	59.80	6858.08	22.37	14.31
	III	4.34	39.92	11.24	97.38	15.36	16.95	48.55	94.31	17.83	54.42	10.73	54.17	57.16	6855.42	19.71	11.65
	IV	6.23	41.82	13.13	99.18	16.11	17.71	49.32	95.06	17.74	54.31	10.63	54.05	52.84	6851.13	15.41	7.36
III	I	3.41	38.97	10.30	96.51	13.45	15.06	46.67	92.33	18.54	55.12	11.41	54.87	60.02	6858.31	22.61	14.53
	II	4.02	39.58	10.91	97.08	11.54	13.16	44.78	90.56	19.06	55.62	11.94	55.38	60.00	6858.28	22.58	14.52
	III	5.08	40.67	11.98	98.07	14.99	16.58	48.16	93.92	16.40	52.95	9.28	52.70	57.52	6855.70	19.98	11.91
	IV	5.95	41.53	12.85	98.93	12.37	13.99	45.63	91.39	19.24	55.82	12.14	55.57	55.25	6853.56	17.85	9.79
IV	I	4.06	39.63	10.46	97.12	13.24	14.83	46.42	92.21	17.59	54.19	12.47	53.89	59.85	6858.11	22.41	14.35
	II	3.57	39.11	10.47	96.62	12.59	14.20	45.78	91.60	18.98	55.54	11.86	55.29	60.05	6858.34	22.64	14.57
	III	4.08	39.64	10.97	97.14	13.97	15.57	47.19	92.92	16.95	53.49	9.83	53.24	59.78	6858.05	22.34	14.27
	IV	6.02	41.62	12.93	98.88	16.20	17.80	49.40	95.15	18.62	55.22	11.52	54.97	52.36	6843.77	14.91	6.85

EQUILIBRIUM DISTRIBUTION

I				II				III				IV			
I	II	III	IV	I	II	III	IV	I	II	III	IV	I	II	III	IV
.30	.02	.08	.01	.07	.09	.02	.01	.05	.02	.11	.02	.02	.00	.04	.14

NOTES

1. Portions of this paper were presented at "Behavioral
Models in Geography," a session at the 1968 AAG meetings
organized by Kevin R. Cox and Reginald G. Golledge of the
Department of Geography, Ohio State University. The discus-
sant for that presentation was Gunnar Olsson of the Department
of Geography, University of Michigan. His comments and those
of Eric G. Moore (Northwestern University) and Kevin R. Cox
have been very useful in preparing this manuscript. Their
assistance is appreciated.

2. A basic discussion on place utility and its role in
intra-urban migration processes can be found in Julian Wolpert,
"Behavioral Aspects of the Decision to Migrate," *Papers of the
Regional Science Association,* Vol. 15 (1965), 159-169.
Further elaboration in an operational context appears in
Lawrence A. Brown and David B. Longbrake, "Migration Flows in
Intra-urban Space: An Ecological Approach to Place Utility,"
(Departments of Geography, Ohio State University and Univer-
sity of Iowa, 1968 (Xerox)).

3. Other papers in this series include Lawrence A. Brown
and Eric G. Moore, "Intra-urban Migration: An Actor Oriented
Framework," (Departments of Geography, Ohio State University
and Northwestern University, 1968 (Xerox)); Lawrence A. Brown
and David B. Longbrake, *op. cit.;* Lawrence A. Brown and Frank
E. Horton, "Functional Distance: A Note on an Operational
Approach," (Departments of Geography, Ohio State University
and The University of Iowa, 1968 (Xerox); Lawrence A. Brown
and Frank E. Horton, "Social Area Change: An Empirical
Analysis," (Departments of Geography, Ohio State University and
The University of Iowa, 1969 (Xerox)); and Lawrence A. Brown
and Frank E. Horton, "Migration Flows in Intra-urban Space:
A Markov Chain Approach," (Departments of Geography, Ohio
State University and The University of Iowa), in preparation.
In addition current research efforts are being directed towards
the analysis of search patterns of households seeking a new
residential site, spatial patterns of acquaintance circles of
recent intra-urban migrants, and sequential residential locations of households within a single urban area from 1950 to
1960 ("migrant life lines"). Most of the data for these

3. (contd.) ongoing efforts results from a survey carried on in Cedar Rapids, Iowa in the spring of 1967.

4. The pitfalls of the ecological approach are well discussed in the literature. See, for example, Hubert M. Blalock, *Causal Inferences in Non-experimental Research* (Chapel Hill: University of North Carolina Press, 1964); L.A. Goodman, "Ecological Regression and Behavior of Individuals," *American Sociological Review*, Vol. 18 (1953), 663-664; and W.S.Robinson, "Ecological Correlations and the Behavior of Individuals," *American Sociological Review*, Vol. 15 (1950), 351-357. Generally, however, arguments against ecological approaches point to its uses for inferences about individual behavior. By contrast, the use of ecological approaches in urban process models, given the restrictions imposed by data availability, is generally accepted. See, for example, Ira S. Lowry, "A Model of Metropolis," Report Number RM-4035-RC, (The Rand Corporation), and Ira S. Lowry, "Seven Models of Urban Development: A Structural Comparison," Report Number P-3673, (The Rand Corporation).

5. This is reported in Lawrence A. Brown and David B. Longbrake, *op. cit.* Evaluation of the results of that study proceeded on two bases. On the one hand the functions, and the findings implied by detailed analysis of the functions, were deemed satisfactory because of their consistency with the body of theory and empirical findings relevant to intra-urban migration processes. In terms of their intended use for operational models of intra-urban migration systems, however, that approach was found lacking, and several alternatives were suggested.

6. Lawrence A. Brown and Eric G. Moore, *op. cit.*

7. For a discussion of comparative place utility in a general context, see Lawrence A. Brown, *Diffusion Processes and Location: A Conceptual Framework and Bibliography* (Philadelphia: Regional Science Research Institute, Bibliography Series Number 4, 1968).

8. In order to tie in directly with our previous work on place utility (Brown and Longbrake, *op.cit.*), the term origin-destination (O-D) zone, a unit used in traffic analyses, will be employed rather than terms such as "areal unit" or "census tract."

193

9. Consistent with previous work by the authors, the term environmental characteristics of a household will be taken to refer to the composite of its neighborhood, dwelling unit, and relative location vis-a-vis other nodes in urban space.

10. Brown and Longbrake, *op. cit.*; Peter H. Rossi, *Why Families Move: A Study in the Social Psychology of Urban Residential Mobility* (New York: The Free Press, 1955); James W. Simmons, "Changing Residence in The City: A Review of Intra-Urban Mobility," *Geographical Review*, Vol. 58 (1968), 622-651.

11. It is conceivable that equation (2) should also include a lag factor to account for the time lapsing between the occurrence of a vacancy and that vacancy being ready for the market. It is thought, however, that this factor is not of sufficient length to be significant in terms of the model presented here.

12. Brown and Moore, *op. cit.*, do not provide a detailed treatment of the mechanisms by which vacancies are generated. For a treatment of this aspect of the intra-urban migration process, particularly as regards the development of new residential sites, the following may be consulted: F. Stuart Chapin and Shirley F. Weiss, *Factors Influencing Land Development* (Chapel Hill: Institute for Research in Social Science, University of North Carolina, 1962); Edward J. Kaiser, "Location Decision Factors in a Producer Model of Residential Development," *Land Economics*, Vol. 44 (1968), 351-362; Shirley F. Weiss, John E. Smith, Edward J. Kaiser, and Kenneth B. Kenney, *Residential Developer Decisions* (Chapel Hill: Institute for Research in Social Science, University of North Carolina, 1966); and Edward J. Kaiser, *A Producer Model for Residential Growth* (Chapel Hill: Institute for Research in Social Science, University of North Carolina, 1968).

13. For examples of this approach see Brian J.L. Berry and Robert A. Murdie, *Socio Economic Correlates of Housing Condition* (Toronto, Urban Renewal Study, Metropolitan Toronto Planning Board, 1965); Robert A. Murdie, *Factorial Ecology of Metropolitan Toronto, 1951-1961: An Essay on the Social Geography of the City* (Chicago, Research Paper No. 116, Department of Geography, University of Chicago, forthcoming); and Philip H. Rees, "The Factorial Ecology of Metropolitan Chicago, 1960 " M.A. Thesis, (Department of Geography, University of Chicago, 1968)

14. For a discussion of linear programming and its application in an economic type framework, see Robert Dorfman, Paul A. Samuelson, and Robert M. Solow, *Linear Programming and Economic Analysis* (New York: McGraw-Hill, 1958); Walter W. Garvin, *Introduction to Linear Programming* (New York: McGraw-Hill, 1960); and David Gale, *The Theory of Linear Economic Models* (New York: McGraw-Hill, 1960).

15. On the surface it appears that a linear programming model of the type used here implies an optimization for the system, rather than for each individual within the system. In addition to the welfare and social control aspects of such an implication, this observation would seem to throw doubt upon the validity of employing a linear programming format for the situation of intra-urban migration. However, in the restrictive setting of a highly competitive and fragmented market situation, where individual households are operating to maximize gain in experienced place utility, it is believed that the competitive equilibrium resulting from individual household efforts would be identical to the equilibrium allocation resulting from optimizing system-wide efficiency criteria. Discussion relevant to this point can be found in Paul A. Samuelson, "Spatial Price Equilibrium and Linear Programming," *American Economic Review*, Vol. 42 (1952), 283-303 and William J. Baumol, "Activity Analysis in One Lesson," *American Economic Review*, Vol. 48 (1958), 837-873.

16. Duane F. Marble and John D. Nystuen, "An Approach to Direct Measurement of Community Mean Information Fields," *Papers of the Regional Science Association*, Vol. 11 (1963), 99-109. Richard L. Morrill and Forrest R. Pitts, "Marriage, Migration, and the Mean Information Field: A Study in Uniqueness and Generality," *Annals of the Association of American Geographers*, Vol. 57 (1967), 401-422.

17. Lawrence A. Brown and David B. Longbrake, *op. cit.*

18. James M. Beshers and Edward O. Laumann, "Social Distance: A Network Approach," *American Sociological Review*, Vol. 32 (1967), 225-236. For a general discussion of social distance, see Edward O. Laumann, *Prestige and Association in an Urban Community* (New York: Bobbs-Merrill, 1966). A discussion of the applicability of the approach of Beshers and Laumann in a geographic context is to be found in Lawrence A. Brown and Frank E. Horton, "Functional Distance: A Note on an Operational Approach," *op. cit.*

19. There is an obvious scale difference between the measure suggested for a^*_{ijk}, which varies from 0 to 1, and those suggested for d_{ijk}, which range from approximately 1 to 330. In combining these, therefore, a standardization procedure should be carried out. One approach is to convert the measures of d_{ijk} into probabilities, based upon the relative values of the mean first passage time in Table 4.

20. The figures in Table 5, which were estimates from the Cedar Rapids Questionnaire data, should be taken with caution since the sample underlying these figures is relatively small.

21. An earlier version of this paper did not include the k subscript to denote socioeconomic categories of households. It should be noted that including the k subscript does not alter the operation or interpretation of the model. Its prime effect is merely to further subdivide the sample. Thus, each zone is in effect subdivided into five strata, given the employment of the socioeconomic categories in Table 1.

22. Richard H. Day, *Recursive Programming and Production Response* (Amsterdam, Netherlands: North-Holland, 1963); Kenneth J. Schlager, "Recursive Programming Theory of the Residential Land Development Process," *Highway Research Record*, No. 126 (1966).

23. Benjamin H. Stevens, "Linear Programming and Location Rent," *Journal of Regional Science*, Vol. 3 (1961), 15-26.

24. Stevens, *op. cit.*

25. One operational problem exists with respect to the linear programming allocation model that should be pointed out. It appears that application of the model would result in over allocation of migrants to nearby vacancy sites. This would be mitigated somewhat by operating the model recursively using a relatively short time interval as suggested in the text. Furthermore, to the extent that such over allocation is the result of employing a distance decay type function to estimate a^*_{ijk}, the problem will disappear with the use of a higher degree function. However, in a deterministic-type framework such as that proposed here there will always be a tendency to allocate a migrant household to the location indicated by the "peak" of its "cost" curve. This is usually inconsistent with the real world data from which the curve is constructed since, as indicated by the curve itself, that data displays a much wider range of values. To solve this problem, three

25. (contd.) strategies are available. One has already been discussed: that of holding the time interval to a small duration. This would have the effect of spreading the spatial availability of vacancies for any time interval. A second approach is to convert the primal to a Monte Carlo simulation model and to compute the utility measures from the result of implementing it, drawing upon the relationship between place utility and migration costs as pointed out in equations (6) and (7). A third approach is to place the problem in the context of statistical linear programming, although techniques for operating the cost variable in a stochastic framework do not appear perfected. For a discussion see Walter W. Garvin, *op. cit.*, 172-190.

26. Brown and Longbrake, *op. cit.*

27. Richard J. Chorley, "Geography and Analogue Theory," *Annals of the Association of American Geographers*, Vol. 54 (1964), 127-137.

28. For a discussion of alternative means of adjusting to extreme stress resulting from dissatisfaction with one's residence site, see Brown and Moore, *op. cit.*

THE SCALING OF LOCATIONAL PREFERENCES*

Gerard Rushton

Michigan State University

A. Introduction

In studies of spatial choice, many results have had questionable relevance to other areas because they reflected some
of the properties of the particular spatial system from which
they had been derived; the distance-decay function is one
example. However, in other disciplines where models of choice
have been developed, a conscious attempt has been made to construct models of preference from which conclusions can be
derived that are independent of the particular set of

---

*I wish to acknowledge and to thank Mr. Robert Kern, who
was responsible for writing for me the computer program
"REVPREF" which accomplishes all of the analyses described in
this paper. A listing of this program and a description of
input and output forms are available on request to the Director,
Computer Institute for Social Science Research, Michigan State
University, East Lansing, Michigan. I would also like to
acknowledge the use of J.B. Kruskal's program for multi-
dimensional scaling and the work of the technical section of
the Computer Institute for Social Science Research in adapting
it to the C.D.C. 3600 Computer. Acknowledgement is also due
to Mrs. Nancy Hammond for her editorial assistance.

alternatives where choices were observed.   Thus a separation
is made between preferences and opportunities.   Models of
spatial choice should attempt to define a preference-scale
which orders all conceivable alternatives as do many other
models of choice.

The purpose of this paper is, therefore, to describe and
to test a methodology for finding space-preference structures
from data which describe spatial choices.   The methodology
used is the method of paired comparisons, made possible by the
development of a computer algorithm for searching and ordering the spatial-relationship of people to alternatives, and
the method of multi-dimensional scaling.

B. Spatial Choice

Spatial behavior implies a search among alternatives.
The criteria for ultimate choice are presumably relative,
rather than absolute, and therefore the process of spatial
choice hypothesized here is one in which a person compares each
alternative with every other one and selects that which he ex-
pects will give him the greatest satisfaction.   Presumably,
any viable model of spatial choice should, in some sense, imitate this procedure.

If the ultimate criterion of choice in this search is
anticipated satisfaction, then, in constructing a model, the
goal must be to express, for any possible locational choice,
the degree of expected satisfaction of each alternative as a
function of the relevant characteristics of the environment
and of the decision-maker.   In view of present knowledge, this
goal is far away, but a step in that direction would be the
discovery of a preference function that in an ordinal way indicated which of several alternatives would give the greatest
satisfaction.   The extent to which decision-makers have the
same preference function is, of course, not known, and so it

will be necessary to design tests to determine the extent to which the spatial choice patterns of different people are consistent with a single preference function. Tests for this purpose do exist.

The interest, therefore, is in passing from an individual statement of preference to the subjective preference function. For any one individual, it is possible to derive that function providing a large number of independent yet consistent choices have been observed. With spatial choice data, however, it is not normally possible to observe a large number of any individual's choices. In such instances, it is assumed that the decisions of different people are generated from similar preference functions. This assumption may seem rash although it will be shown later in this paper that the measure of dissimilarity between locational types constructed from aggregate data is also a measure of inter-personal consistency.

A few recent studies have come close to establishing subjective preference functions for spatial choice. Gould,[1] for example, has evaluated preferences for nominal locations where students were asked to rank states according to their preferences for residing there. Peterson[2] has evaluated preferences between different attributes which locations possess where subjects made decisions on the quality of neighborhoods for residential purposes. A third type of spatial preference was hypothesized by Christaller[3] where neither the names of locations nor the attributes which locations possessed were deemed relevant to spatial choice; rather, the ordering of preference was derived solely from the distance-relationships between the individual and central places.

A subjective preference function orders (that is, ranks) an individual's preferences for a set of objects. In spatial choice, the objects that require ranking are locations with different amounts of relevant properties and different spatial

relations. In spatial choice, for example among towns in
which a consumer makes purchases, the relevant property is related to such questions as: Does the town offer the good in
question? Does it offer other goods? Does it have several
outlets for the good: These are presumably independent criteria which the consumer applies to each town; but before
seeking operational definitions for each, it may be noted that
each measure is related to the population size of towns.
Much empirical work supports the contention that large towns
are more likely to offer a particular good than small towns,
that they will offer more goods and services, and that they
will have more establishments. Therefore, in this study the
relevant properties of towns are hypothesized to be summarized
in the one variable, town population. The spatial relationship of interest is the separation between the individual and
the town. Although time-distance is probably a more accurate
operational definition of this distance than distance in miles,
in the interests of simplicity and availability, this study
uses distance in miles.

Thus, it is possible to take any spatial alternative
facing an individual out of its unique context and to assign
it to its corresponding locational type. Here locational type
is defined according to its spatial relationships and the quantity of the property town size it possesses. Choices from
among unique spatial alternatives are thereby considered equivalent to choices between locational types. A degree of generality has been introduced that allows a meaningful comparison
to be made of decisions made in unique environmental contexts.

If locational choice can be regarded as a choice from
among alternatives, then oberservations of spatial choice can be
regarded as paired comparison data where the locational type
to which the chosen location belongs is preferred to all locational types to which the rejected alternatives belong. For

this conceptualization of the process of spatial choice, the
only assumption made is that choice among a set of alternatives
is equivalent to choice between all the paired combinations.

C. The Method of Paired Comparisons

In understanding spatial choice, it is surely this unique
ranking of spatial situations which is sought. Whether spatial choice is made by comparing each possibility against such
a mental ranking of all conceivable opportunities and choosing
that which has highest rank, or by comparing each actual alternative against every other and choosing that which gives the
highest expected utility the outcome should be the same. To
imitate either process the ranking of possible spatial situations must first be found and this ranking can be determined
from paired comparison data by techniques developed in psychology.[4] Much of the literature concerning the method of
paired comparisons deals with the simple comparison of separate
things, but Guttman[5] has shown that the principles involved can
be extended to include the comparison of combinations of things.

> The problem of paired comparisons arises when it
> is desired to obtain numerical values for a set
> of n things, with respect to one characteristic,
> such that these values will represent the judge-
> ments of n individuals . . . . The judgements
> vary from person to person (and possibly within
> a person), and the problem is to determine a set
> of numerical values for the things being compared
> that will in some sense best represent or average
> the judgements of the whole population.
>
> In some situations, the things being compared may
> be simple items or objects; this we shall call the
> case of *ordinary* comparisons. In other situations
> the things may be *combinations* of items or objects.[6]

The particular least squares' estimating procedure described by Guttman is inappropriate in practice for problems of
spatial choice, since it requires that all individuals make

judgements on all of the n(n-1)/2 comparisons, but the method
has been developed for situations where judgements on all comparisons are incomplete.[7]

D. Study Objectives

The first objective of this paper is to derive the subjective ranking of locational types from empirical data of
actual spatial choices made for a particular purpose.   The
second is to try to progress beyond the purely ordinal scale,
which a ranking of possibilities gives us, to a cardinal measure of the subjective distance between spatial situations.
A third objective is to test the resulting preference surface
for consistency.

E. The Basic Matrix

The data used describes the towns selected by a random
sample of rural people in Iowa for their major grocery purchases.[8]   A second basic source of data is the location and
sizes of all Iowa towns with a 1960 population greater than
50.   This forms the description of alternative opportunities
among which choices were made by members of the sample.

The effect of assigning any particular town to its corresponding locational type is to take spatial choice and spatial
alternatives out of their unique context.   All towns within
twenty-eight miles of each member of the sample were assigned
to one of thirty locational types.   These types are defined
in Figure 1.   The locational type of the town chosen for
major grocery purchases by the household is then considered
preferred to all other locational types available.   Table 1
shows a portion of this basic matrix.   All further computa-
tions described in this work are from the data in this table.

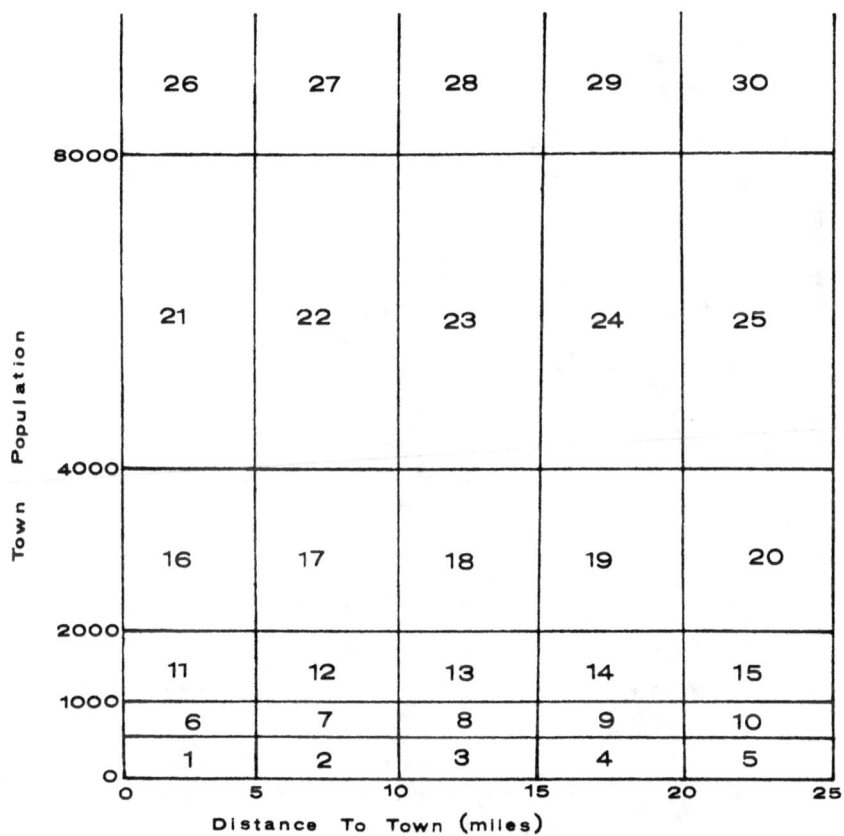

	26	27	28	29	30
8000					
	21	22	23	24	25
4000					
	16	17	18	19	20
2000					
	11	12	13	14	15
1000	6	7	8	9	10
	1	2	3	4	5
0					

Town Population

Distance To Town (miles)

0   5   10   15   20   25

Figure 1   Definition of Locational
Types

TABLE 1

REVEALED SPACE PREFERENCE-RAW DATA MATRIX BY RESPONDENTS

Household ID	\ Locational Type → 1	2	3	4	5	6	7	8	9	10	11	12	13	14	15	16	17	18	19	20	21	22	23	24	25	26	27	28	29	30
1		*	*	*	*				*	*	*		*	*	*															
2		*	*	*	*				*	*			*	*	*		1		1						*					
3	*	*	*	*	*		*	*	*	*	1	1	1	*	*	1		1	*											
3		*	*	*	*		*	*	*	*	1		1	*	*		1	*	*											
4		*	*	*	*				*	*	*	1	*	*	*	1		1												
6		*	*	*	*	*			*	*			*	*											*					
9			*	*	*			*	*	*			*	*	*		*	*	*	*										
10		*	*	*	*			*	*	*		*	*	*	*		*	*	*	*										
11		*	*	*	*				*	*			*	*	*				*	*										
12		*	*	*	*				*	*		*	*	*	*		*		*	*		*	1							
13	1	*	*	*	*				*		*	*	*	*					*	*		*	1							
14		*	1	*	*		1		*		1	1	*	*	*		1	*	*	*		*	1							
15		*	*	*	*	*	1		*	*				*	*				*	*										
16		*	*	*	*		1	*	*					*	*		1	1		*										
18			*	*	*		1	*	*					*	*	*		*	*	*										
19			*	*	*		1	*	*					*	*	*	1		*	*										
20		*	*	*	*		*	*	*			*		*	*			1	*	1										
21		*	*	*	*				*	*		*	*	*			1													
24		*	*	*	*					*	1	1	*	*	*		*		*							*				
25		*	*	*	*	*	*	*		*	1	1	*	*	*		*													
26		*	*	*	*						1																			*
27		*	*	*	*					1	1	1																		
28			*	*	*			*		*	1	1																	1	
29		*	*	*	*														*	*			*							
30			*	*	*		*	*		*	1																			
32	1		*	*	*	*	*			*	1	1	*	*	*															
32	1		*	*	*	*				*			*	*				*				*								
33	1	*	*	*	*				*		*	*	*	*	*		1	*	*	*		1			*				1	
34	1	*	*	*	*			*	*					*	*					*					*			*		
35		*	*	*	*																								1	*

Source: Computed from data described in footnote 8 (Part of table only shown)

1 = Locational Type Patronized
* = Locational Type Rejected
Blank = Locational Type Not Present

F. Perceived Similarity of Locational Types

A measure of the similarity between any two locational types is the degree to which one is preferred by persons who can choose both types. This degree can be computed from Table 1 as the proportion of times one type is chosen over the other when both are present. Table 2 shows the number of times the column locational type was preferred to the row type. The probability that the jth locational type is preferred to the ith locational type P_{jpi} can be computed from Table 2.[9]

Let T_{ij} be the cell total in ith row and jth column of Table 2 and T_{ji} be the cell total in jth row and ith column.

$$P_{jpi} = \frac{T_{ij}}{T_{ij} + T_{ji}}$$

Where $T_{ij} = 0$, and $T_{ji} = 0$, P_{jpi} and P_{ipj} are undefined

Where $T_{ji} = 0$, and $T_{ij} > 0$, $P_{jpi} = 1$

$T_{ij} = 0$, and $T_{ji} > 0$, $P_{jpi} = 0$

Thus, whenever $T_{ji} + T_{ij} > 0$, $P_{jpi} + P_{ipj} = 1$

and $P_{jpi} = 1 - P_{ipj}$

These probabilities are found in Table 3. In this table the order of both the columns and the rows represents an ordering of locational types from the preference data according to the ratio in the last column of this table. Each number in this column shows the proportion of times the locational type for the row was preferred to the other types.

TABLE 2

SEGMENT OF THE REVEALED PREFERENCE DATA MATRIX

Cells Show Number of Times Sample Households Preferred Column Locational Type to Row Type

	1	2	3	4	5	6	7	8	9	10	11	12	13	14	15	16	17	18
1	29.8	9.0	3.0	1.0	0.0	27.0	15.3	3.0	1.0	0.0	23.5	47.0	7.2	0.0	0.0	22.0	33.5	25.9
2	127.7	21.5	2.0	2.0	0.0	90.5	55.3	12.0	0.0	0.0	85.0	94.3	23.5	0.0	0.0	81.0	77.5	65.7
3	180.0	27.5	11.0	0.0	0.0	188.5	97.5	22.0	3.0	0.0	176.5	154.2	25.0	0.0	0.0	119.0	133.0	101.5
4	243.3	36.0	8.0	9.0	0.0	228.0	137.3	14.0	8.0	0.0	211.5	190.5	44.0	0.0	0.0	126.0	193.0	106.8
5	209.8	37.0	8.0	5.0	0.0	223.0	133.3	18.0	8.0	0.0	224.0	193.7	47.5	0.0	0.0	165.0	180.5	100.7
6	5.5	0.0	0.0	0.0	0.0	4.0	5.0	0.0	0.0	0.0	6.5	6.5	1.0	0.0	0.0	2.0	7.5	4.0
7	30.5	5.5	1.0	0.0	0.0	36.5	17.3	1.0	2.0	0.0	14.5	24.0	5.0	0.0	0.0	12.0	20.5	10.0
8	41.7	7.0	1.0	1.0	0.0	41.0	33.0	3.0	1.0	0.0	40.0	31.8	6.2	0.0	0.0	27.0	32.5	19.2
9	66.7	5.0	4.0	0.0	0.0	63.0	34.0	2.0	0.0	0.0	51.0	54.3	12.7	0.0	0.0	45.0	54.5	15.7
10	65.8	10.5	2.0	2.0	0.0	51.5	34.8	5.0	0.0	0.0	65.5	56.8	10.0	0.0	0.0	36.5	53.0	27.8
11	2.5	0.0	0.0	0.0	0.0	1.0	0.0	0.0	0.0	0.0	1.0	0.0	1.5	0.0	0.0	0.0	4.0	1.5
12	8.5	3.0	0.0	0.0	0.0	10.0	3.0	0.0	1.0	0.0	5.0	8.3	1.0	0.0	0.0	8.0	10.0	4.0
13	23.3	1.5	0.0	1.0	0.0	24.5	17.0	0.0	1.0	0.0	24.5	18.5	7.5	0.0	0.0	8.0	23.0	10.3
14	32.8	6.0	0.0	1.0	0.0	46.5	16.5	2.0	1.0	0.0	29.5	28.3	3.8	0.0	0.0	17.5	17.0	9.8
15	43.8	3.0	1.0	1.0	0.0	35.0	15.3	0.0	0.0	0.0	27.0	37.3	9.5	0.0	0.0	23.5	22.0	11.3
16	0.0	0.0	0.0	0.0	0.0	0.0	0.0	0.0	0.0	0.0	1.0	0.0	0.0	0.0	0.0	2.0	0.0	1.0
17	3.0	0.5	0.0	0.0	0.0	8.0	4.5	1.0	0.0	0.0	6.0	6.0	1.0	0.0	0.0	1.0	6.5	1.0
18	16.6	3.0	2.0	0.0	0.0	13.5	6.3	1.0	2.0	0.0	7.5	3.0	2.3	0.0	0.0	3.0	3.5	4.5
19	16.5	0.5	3.0	1.0	0.0	25.0	10.0	2.0	0.0	0.0	20.5	19.0	3.0	0.0	0.0	11.0	17.5	3.5
20	9.3	4.5	1.0	0.0	0.0	12.0	10.3	2.0	0.0	0.0	17.5	20.0	2.5	0.0	0.0	6.0	16.0	5.8
21	0.0	0.0	0.0	0.0	0.0	2.0	0.0	0.0	0.0	0.0	0.0	0.0	1.0	0.0	0.0	0.0	0.0	0.0
22	3.0	0.0	0.0	0.0	0.0	6.0	2.0	0.0	0.0	0.0	1.0	2.0	1.0	0.0	0.0	1.0	1.0	1.0
23	13.0	1.5	1.0	0.0	0.0	5.0	5.0	0.0	0.0	0.0	8.0	5.0	2.0	0.0	0.0	5.0	1.5	2.0
24	16.0	2.0	1.0	0.0	0.0	8.5	6.5	1.0	0.0	0.0	13.5	8.0	2.0	0.0	0.0	9.0	7.5	2.0
25	14.3	2.0	0.0	0.0	0.0	11.0	8.3	3.0	0.0	0.0	16.5	12.0	2.0	0.0	0.0	10.0	7.5	3.8
26	0.0	0.0	0.0	0.0	0.0	0.0	0.0	0.0	0.0	0.0	0.0	0.0	0.0	0.0	0.0	2.0	1.0	1.5
27	2.0	0.0	0.0	1.0	0.0	1.0	1.0	0.0	1.0	0.0	2.0	1.0	0.0	0.0	0.0	0.0	0.0	0.0
28	2.0	1.0	0.0	0.0	0.0	5.5	3.0	0.0	0.0	0.0	4.0	2.0	1.0	0.0	0.0	1.0	0.0	0.0
29	7.3	0.0	0.0	0.0	0.0	13.0	7.3	0.0	0.0	0.0	8.5	5.0	0.0	0.0	0.0	6.0	6.0	2.8
30	13.3	1.0	0.0	0.0	0.0	4.5	5.3	0.0	0.0	0.0	9.0	9.3	2.0	0.0	0.0	9.5	4.0	3.8

Source: Computed from data described in footnote 8 (part of table only shown)

TABLE 3

PROBABILITY THAT COLUMN LOCATIONAL TYPE IS PREFERRED TO ROW TYPE

	21	16	27	11	26	28	22	23	17	29	6	12	18	24	1	7
22	-1.00	-1.00	-1.00	-1.00	-1.00	-1.00	-1.00	0.00	-1.00	-1.00	0.25	0.00	0.00	0.00	0.00	0.00
16	-1.00	-1.00	-1.00	1.00	0.00	0.00	0.00	0.17	0.00	0.14	0.00	0.00	0.25	0.00	0.00	0.00
27	-1.00	-1.00	-1.00	0.40	1.00	0.25	0.00	0.00	0.00	0.00	0.11	0.20	0.00	0.00	0.06	0.05
11	-1.00	0.00	0.60	-1.00	-1.00	0.56	0.00	0.11	0.40	0.35	0.20	0.00	0.17	0.04	0.10	0.00
26	-1.00	1.00	0.00	0.44	-1.00	0.00	0.00	0.00	0.50	-1.00	0.00	0.00	0.50	0.00	0.00	0.00
28	-1.00	1.00	0.75	1.00	1.00	-1.00	1.00	0.67	0.13	0.00	0.42	0.22	0.10	0.06	0.05	0.10
22	-1.00	0.83	1.00	0.89	1.00	0.33	-1.00	0.00	1.00	0.25	0.35	0.17	0.00	0.06	0.07	0.09
23	1.00	1.00	1.00	0.60	1.00	0.88	1.00	-1.00	0.38	0.57	0.37	0.27	0.08	0.27	0.24	0.17
17	-1.00	0.86	1.00	0.65	0.50	1.00	0.00	0.63	-1.00	0.00	0.52	0.38	0.22	0.06	0.08	0.18
29	-1.00	1.00	0.89	0.80	-1.00	0.58	0.75	0.43	1.00	-1.00	0.81	0.45	0.29	0.00	0.30	0.38
6	0.75	1.00	0.80	1.00	1.00	0.78	0.65	0.63	0.48	0.19	-1.00	0.39	0.23	0.32	0.17	0.12
12	1.00	0.75	1.00	0.83	0.50	0.90	0.83	0.73	0.63	0.55	0.61	-1.00	0.57	0.34	0.15	0.11
18	1.00	1.00	1.00	0.96	1.00	1.00	1.00	0.92	0.78	0.71	0.77	0.43	-1.00	0.71	0.39	0.38
24	1.00	1.00	0.94	0.90	1.00	0.95	0.94	0.73	0.94	1.00	0.68	0.66	0.29	-1.00	0.50	0.39
1	1.00	1.00	0.95	1.00	1.00	0.90	0.93	0.76	0.92	0.70	0.83	0.85	0.61	0.50	-1.00	0.33
7	0.92	1.00	1.00	0.94	1.00	0.93	0.91	0.83	0.82	0.62	0.88	0.89	0.62	0.61	0.67	-1.00
13	1.00	1.00	1.00	1.00	1.00	0.88	0.97	0.88	0.96	1.00	0.96	0.95	0.82	0.67	0.77	0.77
30	1.00	1.00	1.00	1.00	1.00	0.90	1.00	0.81	0.73	0.73	1.00	0.78	0.71	0.57	0.75	0.78
25	1.00	1.00	1.00	1.00	1.00	1.00	1.00	1.00	1.00	1.00	0.92	0.92	1.00	0.67	0.83	0.73
20	1.00	1.00	1.00	1.00	1.00	1.00	1.00	1.00	1.00	0.85	0.92	1.00	0.85	1.00	0.90	0.77
8	1.00	1.00	1.00	1.00	1.00	0.99	1.00	1.00	0.97	1.00	1.00	1.00	0.95	0.86	0.93	0.97
2	1.00	1.00	1.00	0.95	1.00	1.00	1.00	0.99	0.99	1.00	1.00	0.97	0.96	0.94	0.93	0.91
19	1.00	1.00	1.00	1.00	1.00	1.00	1.00	1.00	1.00	1.00	0.96	1.00	1.00	-1.00	0.85	0.91
3	1.00	1.00	1.00	1.00	1.00	1.00	1.00	0.99	1.00	1.00	0.99	1.00	0.98	0.98	0.98	0.99
4	1.00	1.00	0.99	1.00	1.00	1.00	1.00	1.00	1.00	1.00	1.00	1.00	0.99	1.00	1.00	1.00
9	1.00	1.00	0.98	1.00	1.00	1.00	1.00	1.00	1.00	1.00	1.00	1.00	0.89	1.00	0.99	0.94
15	1.00	1.00	1.00	1.00	1.00	1.00	1.00	1.00	1.00	1.00	1.00	1.00	1.00	1.00	1.00	1.00
14	1.00	1.00	1.00	1.00	1.00	1.00	1.00	1.00	1.00	1.00	1.00	1.00	1.00	1.00	1.00	1.00
10	1.00	1.00	1.00	1.00	1.00	1.00	1.00	1.00	1.00	1.00	1.00	1.00	1.00	1.00	1.00	1.00
5	1.00	1.00	1.00	1.00	1.00	1.00	1.00	1.00	1.00	1.00	1.00	1.00	1.00	1.00	1.00	1.00

Source: Computed from Table 2 (-1.0 is missing data)

208

TABLE 3 (Continued)

PROBABILITY THAT COLUMN LOCATIONAL TYPE IS PREFERRED TO ROW TYPE

	13	30	25	20	8	2	19	3	4	9	15	14	10	5	Per Cent ≥ .5
21	0.08	0.00	0.00	0.00	0.00	0.00	0.00	0.00	0.00	0.00	0.00	0.00	0.00	0.00	100.00
16	0.00	0.00	0.00	0.00	0.00	0.00	0.00	0.00	0.00	0.02	0.00	0.00	0.00	0.00	96.30
27	0.00	0.00	0.00	0.00	0.00	0.00	0.00	0.00	0.01	0.00	0.00	0.00	0.00	0.00	96.30
11	0.06	0.00	0.00	0.00	0.00	0.00	0.05	0.00	0.00	0.00	0.00	0.00	0.00	0.00	92.59
26	0.00	0.13	0.10	0.00	0.00	0.01	0.00	0.00	0.00	0.00	0.00	0.00	0.00	0.00	88.46
28	0.07	0.00	0.00	0.00	0.00	0.01	0.00	0.00	0.00	0.00	0.00	0.00	0.00	0.00	82.14
22	0.03	0.19	0.00	0.00	0.03	0.01	0.00	0.00	0.00	0.00	0.00	0.00	0.00	0.00	82.14
23	0.13	0.27	0.00	0.15	0.00	0.01	0.00	0.01	0.00	0.00	0.00	0.00	0.00	0.00	75.86
17	0.04	0.27	0.00	0.08	0.00	0.00	0.00	0.00	0.00	0.00	0.00	0.00	0.00	0.00	75.00
29	0.00	0.00	0.08	0.00	0.05	0.03	0.04	0.01	0.00	0.11	0.00	0.00	0.00	0.00	74.07
6	0.04	0.22	0.08	0.15	0.14	0.04	0.00	0.00	0.00	0.01	0.00	0.00	0.00	0.00	72.41
12	0.05	0.29	0.08	0.10	0.07	0.06	0.00	0.00	0.00	0.06	0.00	0.00	0.00	0.00	58.62
18	0.18	0.43	0.00	0.23	0.03	0.07	0.00	0.02	0.01	0.07	0.00	0.00	-1.00	0.00	58.62
24	0.33	0.25	0.33	0.00	0.00	0.09	-1.00	0.02	0.00	0.00	0.00	0.00	0.00	0.00	53.57
1	0.23	0.22	0.17	-1.00	0.38	0.06	0.15	0.02	0.00	0.00	0.00	0.00	0.00	0.00	51.72
7	0.23	0.20	0.27	-1.00	0.40	0.11	0.09	0.01	0.00	0.33	0.00	0.00	0.00	0.00	48.28
13	-1.00	-1.00	0.00	-1.00	-1.00	0.50	0.00	0.00	0.02	0.00	0.00	0.00	0.00	0.00	44.83
30	0.80	-1.00	0.67	0.60	0.63	0.60	0.00	0.00	0.00	0.43	0.00	0.00	0.00	0.00	35.71
25	1.00	-1.00	-1.00	0.40	0.67	0.37	0.00	0.00	0.00	1.00	0.00	0.00	0.00	0.00	33.33
20	1.00	1.00	-1.00	0.90	0.96	-1.00	0.33	0.10	0.00	-1.00	0.00	0.00	-1.00	0.00	30.77
8	1.00	0.89	0.63	1.00	0.93	0.14	0.33	0.04	0.07	-1.00	0.00	0.00	0.00	0.00	28.57
2	0.94	1.00	0.50	1.00	0.67	0.93	0.86	0.07	0.05	1.00	0.00	0.00	0.00	0.00	27.59
19	1.00	1.00	1.00	1.00	-1.00	0.95	-1.00	0.25	0.00	-1.00	-1.00	-1.00	-1.00	0.00	23.08
3	1.00	1.00	1.00	1.00	1.00	1.00	0.75	-1.00	-1.00	1.00	0.00	0.00	0.00	0.00	14.81
4	0.98	1.00	1.00	1.00	1.00	1.00	1.00	1.00	0.00	-1.00	-1.00	-1.00	-1.00	-1.00	13.79
9	0.93	1.00	1.00	1.00	1.00	1.00	1.00	0.57	-1.00	1.00	-1.00	-1.00	0.00	-1.00	11.11
15	1.00	1.00	-1.00	1.00	1.00	1.00	1.00	-1.00	1.00	1.00	0.00	0.00	-1.00	-1.00	0.00
14	1.00	1.00	1.00	1.00	1.00	1.00	-1.00	1.00	1.00	1.00	-1.00	-1.00	-1.00	-1.00	0.00
10	1.00	1.00	1.00	-1.00	1.00	1.00	-1.00	-1.00	1.00	1.00	-1.00	-1.00	-1.00	-1.00	0.00
5	1.00	1.00	1.00	1.00	1.00	1.00	1.00	1.00	1.00	-1.00	-1.00	-1.00	-1.00	-1.00	0.00

Source: Computed from Table 2 (-1.0 is missing data)

Maximum perceived similarity between any two locational
types, i and j, would have the value 0.5 in Table 3; that is:
$P_{jpi} = 0.5$ and $P_{ipj} = 0.5$. Any departure from this value in
either direction represents an increase in perceived dissimilarity between the locational types in question. A measure of
perceived similarity d_{ij} (distance in perceived dissimilarity
between locational types i and j), is therefore given by:

$$d_{ij} = \left| P_{jpi} - 0.5 \right|$$

G. Scaling the Similarity Measures

From these proximity measures a scale is to be constructed
on which all locational types will be positioned. Interpoint
distances measured from this scale ought to correspond to the
original proximity measures. Such a scale would, of course,
order the locational types and thus would demonstrate that the
spatial choice data can be regarded as having been generated
from a preference structure, since preference structures are an
ordering of all conceivable alternatives. The particular
preference structure that generated the data set would thus
have been identified. Knowing that from it a large proportion
of spatial choices can be predicted would confirm the conceptualization of the spatial choice problem outlined earlier.
From the scale, measures can be made of the dissimilarity between locational types for which no data were present in the
original matrix of interpoint dissimilarities.[10] This possibility has far-reaching implications for the analysis of
choices in an area where the distribution of alternative opportunities differs substantially from that of the area where the
sample data were gathered. Another reason for searching for
a scale is that in the conceptualization of the decision-making
process for spatial choice, the conceptualization is of a rule
being present whereby all alternative spatial opportunities can

be judged and choice made. Such a rule would imply that considerably more than ordinal information on the location of points in a space exists for otherwise the rank order of distances between all conceivable points could not be known. Hence spatial choice appears to be an example of a situation where "*ordinal* information on distances does imply a considerable amount of *interval* information on the location of the points."[11]

The technique described below is based on a ranking of the measures of dissimilarity. Thus qualities of additivity to the measures of dissimilarity are not necessarily imputed. Since they are derived from ordinal relationships (from the paired comparisons) an intrinsic interval level of measurement in the data matrix cannot be assumed even though it is assumed that such a scale must have been present for the matrix to be generated. However, Shepard's work has shown the possibility, in certain specified circumstances, of deriving from ordinal-type data approximate metric scales that have a high degree of accuracy.

Let δ_{ij} be any one of the measures of dissimilarity, described above.[12] For a matrix of such dissimilarities the intent is to represent the n locational types as n points in t dimensional space the interpoint distances of which (d_{ij}) correspond to the observed degree of dissimilarity between the n locational types. Perfect correspondence would mean, for example, that, if locational type i is more similar to type j than it is to type k, then the corresponding interpoint distances would satisfy the same relationship -- for all i, j, k. That is, where $\delta_{ij} > \delta_{ik}$: $d_{ij} > d_{ik}$. The simplest example would be one in which locational types could be so arranged in one-dimensional space that the ranking of interpoint distances corresponded to the ranking of dissimilarities in the probability matrix.

With locational types, assuming complete data, there are n(n-1)/2 dissimilarities. Ignoring, for the moment, the possibilities of time, the dissimilarities can be ranked in ascending order.

$$\delta_{i1j1} <\delta_{i2j2} <\delta_{i3j3} < \cdots \cdots \delta_{iMjM} \cdots \qquad (1)$$

Here M = n(n-1)/2. Let the locational types be called x_i, ... ,x_n and expressed in orthogonal coordinates by $x_i = (x_{i1}, \cdots ,x_{i3}, \cdots ,x_{it})$. Let d_{jk} denote the distance from x_j to x_k; then

$$d_{jk} = \left[\sum_{r=1}^{n} (P_{rj} - P_{rk})^2 \right]$$

Where   j,k are indices for any two points
r is an index for axes
n is the number of orthogonal axes
P_{rj} refers to the projection of point j on axis r

Correspondence between the interpoint distances and the dissimilarities would mean that if the distances were ranked from smallest to largest then the same order of the locational types in (1) above is maintained.

That is:

$$d_{i1ji} <d_{i2j2} <d_{i3j3} < \cdots \cdots \cdots d_{iMjM}$$

In other words, if the locational types are shown on a scatter plot in which the ordinate is dissimilarity (δ) and the abscissa is distance (d), then as the points are traced one by one from bottom to top, the move is always to the right, never to the left. When this requirement has been met, a monotone relationship between dissimilarity and distance has been found.

With empirical data, however, perfect correspondence is rarely achieved and so some measure of goodness of fit is required. Kruskal proposes a test similar to the correlation coefficient that he calls stress.

$$S = \sqrt{\frac{\sum_{i<j} (d_{ij} - \hat{d}_{ij})^2}{\sum_{i<j} d_{ij}^2}} \qquad (2)$$

Where $\hat{d}_{ij}$ is the minimum distance between x_i and x_j that will satisfy the monotonic relationship.

Kruskal has developed an algorithm for finding the orthogonal coordinates for the n points such that for any number of dimensions stress is minimized. When some of the dissimilarities are missing the terms that correspond to the missing dissimilarities are dropped in both the numerator and the denominator of the definition of stress. The extent to which spatial choice can be understood as the application of a subjective ranking of locational types to the particular unique set of spatial alternatives facing an individual is described by the amount of stress on the first dimension. Table 4 shows the computed scale values on the first dimension. In Figure 2, the scale is shown as isopleths of this data. These interpolated lines are of equal scale value on the two variables, population size and distance, which were used in defining the locational types. This surface is called an indifference surface of spatial choice, because it has all of the typical features of a preference surface. The inference to be made is that a person would be indifferent between (that is, expect equal satisfaction from) any two towns placed along one of the isolines, and would prefer (that is, expect most satisfaction) from the town that lies on the highest point of the surface.

TABLE 4

SCALE VALUES FOR THE LOCATIONAL TYPES

Rank	Locational Type	Scale Value
1	21	-1.311
2	16	-1.523
3	27	-1.382
4	11	-0.991
5	26	-1.624
6	28	-0.876
7	22	-1.160
8	23	-0.590
9	17	-0.719
10	29	-0.668
11	6	-0.725
12	12	-0.481
13	18	-0.321
14	24	-0.020
15	1	-0.163
16	7	-0.075
17	13	-0.217
18	30	-0.009
19	25	0.378
20	20	0.532
21	8	0.752
22	2	0.551
23	19	0.812
24	3	1.114
25	4	1.174
26	9	1.266
27	15	1.438
28	14	1.596
29	10	1.517
30	5	1.725

Stress = 0.434

Source: Computed according to algorithm described by J.B. Kruskal, *op. cit.* (1964), 115-129.

One feature of the scale developed is that it is frequently not consistent with the replies of respondents with
respect to individual pairs of locational types.   For example,
although type 16 is shown to give more satisfaction than type
11 on the final scale (Fig. 2) Table 3 shows that on the occasions when a person made a choice between types 11 and 16, type
11 was always preferred.   However, Table 2 shows that there
was only one such occasion and it is therefore not surprising
that in such circumstances sampling error is large.   Indeed,
opportunities for comparison between locational types that are
close on the satisfaction scale might be uncommon, and accordingly, large sampling variations will surround any values
that do occur.   One advantage of the scaling method employed
here is that it uses information in addition to the comparison
of any two locational types to determine the distance between
them on the final scale.   In fact, the analysis has shown that,
given sufficient qualitative statements of the form "locational
type i is preferred to locational types j," it is possible to
derive a metric scale that gives a meaningful distance measure
between any two locational types.   The conclusion might seem
to contradict a time-honored distinction between ordinal and
interval measurement scales; but recent theoretical and experimental results using both the Kruskal scaling technique
used in this paper and other techniques have shown that, given
a sufficient number of inequalities on the interpoint distances,
the location of points in any given number of dimensions are
free to move only within narrow limits before some interpoint
inequality is broken.[13]   Thus in certain circumstances, metric
scales with a very high degree of accuracy can be developed
from purely ordinal data.

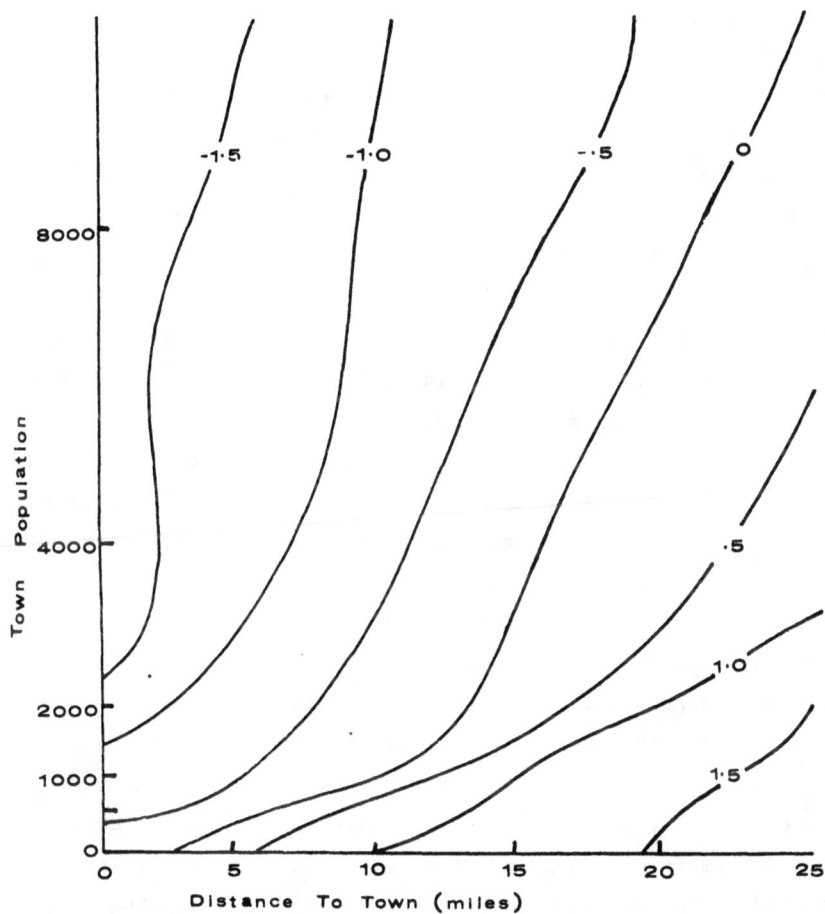

Figure 2.   Space Preference Structure for Grocery
             Purchases :   Iowa 1960

H. Transitivity and the Scale

A more conventional test of whether this assumption of
unidimensionality in the judgement between locational types is
justified, is the extent to which the matrix of probabilities
is fully transitive.   This test is commonly referred to in the
psychological literature as the test for weak stochastic
transitivity.   In Table 5, the data from Table 3 is transformed into a binary matrix by substituting a 1 whenever
$P_{jpi} > .5$ and a 0 whenever $P_{jpi} < .5$.

The coefficient of consistency for such a matrix varies
from 0 (maximum inconsistency) to 1 (complete transitivity).
The 1 would indicate that a unidimensional preference structure
obtains.[14]   The coefficient can be described verbally as one
minus the ratio between the observed number of cyclic triplets
in the matrix to the maximum number of possible cyclic triplets
in the matrix, where a cyclic triplet is an intransitivity.
An intransitivity can be described as the example in which
locational type i is preferred to j, type j is preferred to k,
and, yet type k is revealed preferred to i.   Clearly, the
greater the proportion of such cyclic triplets present of all
such triplets that could be present, the less one can speak
of a unidimensional general preference structure.   The lower
the coefficient of consistency the more likely that several
distinct preference structures exist in the sample population
studied.   The coefficient of consistency[15] was computed from
Table 5.   It has a value of 0.985.

I. Some Unsolved Problems

The Problem of Aggregation

To the extent that the proximity measures are measures of
how frequently the population of individuals disagreed about a
common scale, they provide evidence on the important subject of
inter-personal consistency of spatial behavior.   Where, for

TABLE 5

TRANSITIVITY TEST FOR CONSISTENCY OF PREFERENCE SURFACE

Locational Types

	1 5	1 0	1 4	1 5	9	4	3	1 9	2	8	2 0	2 5	3 0	1 3	7	1 2	1 8	6	1 2	2 9	1 7	2 3	2 2	2 8	2 6	2 1	2 7	1 6	2 1
21	0	0	0	0	0	0	0	0	0	0	0	0	0	0	0	0	0	0	0	2	2	0	2	2	2	2	2	2	0
16	0	0	0	0	0	0	0	0	0	0	0	0	0	0	0	0	0	0	0	0	0	0	2	0	2	2	0	2	2
27	0	0	0	0	0	0	0	0	0	0	0	0	0	0	0	0	0	0	0	0	2	0	2	0	1	0	1	1	1

example, location type A is always preferred to type B, the
scale that represents A higher than B is common to all individuals for that comparison. Where A was preferred to B as
frequently as B was preferred to A, the two points occupy the
same position on the common scale although on the individuals'
scales, A and B may well be separated. Coombs points the way
to future research here when he writes:

> An experimentor may be interested, not in con-
> structing a common stimulus scale for a population
> of individuals, but rather in partitioning a popu-
> lation into subgroups, each of which has a common
> but distinctive stimulus scale. Then the intent
> is to find the distinguishing characteristics of
> the subpopulations which are associated with the
> corresponding distinctions between stimulus scales.[16]

The Problem of Temporal Changes

In addition to the question of how similar the preference
structure of one person is to that of another is the question
of how stable a person's preference structure is through time.
Since the individual operates in a constantly changing spatial
environment one aspect of spatial preference is presumably the
rules by which a person acquires knowledge about the environment and, indeed, anticipates changes in it. Curry and
others[17] have recently emphasized the interdependence, and
consequently the mutual adaptation, of the activities of consumers and retailers in a continuous learning process.
Models are needed

> to separate out the momentary fluctuations in
> behavior which may obscure the behavior's more
> persistent and organized character.[18]

The problem of aggregation and of temporal change can be
shown as the combination of two response matrices. The first
one, Table 6, illustrates the possibility of different individuals, responding differently to the same set of stimuli
(locational types) on a single occasion while the second matrix

refers to responses by the same individual on different
occasions. Combining the two a box can be conceived of with
responses related to the three axes of locational types, individuals and occasions. The spatial learning studies referred
to above become, in this context, a study of individuals'
patterns of responses among occasions with a view to determining whether revealed inconsistencies in a unitary preference
structure by the same individual are the momentary fluctuations
that Coombs refers to or whether they represent a systematic
change in the preference structure itself. That spatial
choice patterns change through time is neither sufficient nor
necessary to prove that preference structures have changed.
Changes in spatial choice patterns result from changes in the
distribution of alternatives, changes in the 'action-space' of
individuals through the learning process, as well as changes
in the individual's preference structure. These three influences must be identified and separated for a meaningful
study of spatial behavior. Presumably, 'action-space' is
related to the distribution of alternatives, whereas preference
structures are independent of any particular distribution of
alternatives. How else could spatial choices be made in new
environments? However, the question of how preference structures change through time, how stable they are through time,
still remains.

The Problem of Surrogates

The operational definition of locational types used in
this study was influenced by the availability of data, but the
conceptualization of the scaling problem described here is
independent of these definitions. The problem of surrogates
is the problem of using intermediate entities as operational
definitions of the more fundamental properties that presumably
one is interested in.[19] Houthaker discusses this problem in
relation to indifference maps in economics in which he argues,

TABLE 6

A RESPONSE MATRIX SHOWING RESPONSES TO

VARIOUS LOCATIONAL TYPES BY VARIOUS

INDIVIDUALS ON A SINGLE OCCASION

	Locational Types				
Individuals	L_1	L_2	L_3	L_j	L_n
I_1	R_{11}	R_{12}		R_{ij}	R_{1n}
I_2	R_{21}	R_{22}		R_{2j}	R_{2n}
I_3	R_{31}	R_{32}		R_{3j}	R_{3n}
$\cdot$ $\cdot$ $\cdot$					
I_i	R_{i1}	R_{i2}		R_{ij}	R_{in}
$\cdot$ $\cdot$					
I_N	R_{N1}	R_{N2}		R_{Nj}	R_{Nn}

TABLE 7

A RESPONSE MATRIX SHOWING RESPONSES TO
VARIOUS LOCATIONAL TYPES ON VARIOUS
OCCASIONS BY THE SAME INDIVIDUAL

Occasions	Locational Types				
	L_1	L_2	L_3	L_j	L_n
O_1	R_{11}	R_{12}	R_{13}	R_{1j}	R_{1n}
O_2	R_{21}			R_{2j}	R_{2n}
O_3	R_{31}			R_{3j}	R_{3n}
' ' '					
O_i	R_{i1}			R_{ij}	R_{in}
' '					
O_n	R_{N1}			R_{Nj}	R_{Nn}

even the quantities described are presumably surrogates for
more abstract properties in several dimensions which each
quantity possesses.[20]   The situation is not uncommon where
it is impossible to derive independent measures of the separate factors that influence behavior; only the order of
their joint effect is known.   The problem then becomes one
of finding the measurement scales both for the factors and
for their effects.   This problem is known as the conjoint
measurement problem and has recently received widespread discussion.[21]

The Problem of Error

     The scaling model described in this paper is a determin-
istic model.   No provision is made for any error or aberrant
response, and since such responses may occur, deviation between actual and model behavior is to be expected.   Unlike
probabilistic type scaling models, such models present problems
in evaluating goodness of fit.[22]   Furthermore, sampling error
will be important when the number of locational types is large
relative to the number of revealed preferences.   Guilford
suggests that "possibly a good rule would be to limit application to where N/n is greater than fifty."[23]

     One writer has recently argued with impressive supporting
evidence that, whenever choice has to be made among alternatives, each of which consists of a number of subjectively
disparate attributes, man's ability "to arrive at an overall
evaluation by weighting and combining or 'trading off' all of
these separate attributes"[24] is not impressive; that is, it
can frequently be shown to be inconsistent with his stated
preferences in evaluating the alternatives on any one of the
subjective attributes.   Shepard describes the results of
experiments that indicate that man often fails in his efforts
to combine correctly the effects of the multitude of factors
that influence his behavior and so fails to choose from the

set of alternatives the one that would give him the greatest satisfaction.

J. Summary

The study of spatial choice should be a study of revealed preference between locational types. As paired comparisons between locational types, such data can be used to establish a preference function from which it is possible to derive the subjective ranking of locational types. By using recent advances in scaling techniques it is possible to derive a measure of the perceived dissimilarity between locational types. Finally, a test for stochastic transitivity in the revealed preference matrix reveals the extent to which the assumption of a unidimensional preference structure is valid.

The purpose of studying spatial behavior should be to derive a description of preferences. Opportunities for spatial interaction are readily observable, and consequently spatial choice is predictable by placing these opportunities against a space preference structure. A model of spatial choice developed within this framework would finally accomplish "the analytical separation of preference and opportunity,"[25] such as was achieved in the economist's model of choice many decades ago. The theory and methods of scaling applied to paired comparison data provide an appropriate analytical structure for solving many geographical problems in which spatial choice is present.

224

NOTES

1. Peter R. Gould, "Structuring Information on Spacio-Temporal Preferences," *Journal of Regional Science*, VII, (Supplement-Winter, 1967), 259-274; Peter R. Gould, "On Mental Maps" (Ann Arbor: Michigan Inter-University Community of Mathematical Geographers, 1966).

2. G.L. Peterson, "A Model of Preference: Quantitative Analysis of the Perception of the Visual Appearance of Residential Neighborhoods," *Journal of Regional Science*, VII (1967), 19-32.

3. W. Christaller, *Central Places in Southern Germany*, trans. by C.W. Baskin (Englewood Cliffs, N.J.: Prentice-Hall, Inc., 1966). Others have drawn attention to this postulate and to its critical role in central place theory: B.J.L.Berry and W. Garrison, "Recent Developments in Central Place Theory," *Papers and Proceedings, Regional Science Association*, IV (1958), 107-120; G. Rushton, R.G. Golledge, and W.A.V. Clark, "Formulation and Test of a Normative Model for the Spatial Allocation of Grocery Expenditures by a Dispersed Population," *Annals, Association of American Geographers*, LVII (1967), 390; G. Rushton, "Analysis of Spatial Behavior by Revealed Space Preference," *Annals, Association of American Geographers*, LIX (1969) forthcoming.

4. Clyde H. Coombs, *A Theory of Data* (New York: John Wiley & Sons, 1964) (especially part 4: "Stimuli Comparison Data"); H.A. David, *The Method of Paired Comparisons* (New York: Hafner Publishing Company, 1963); J.P. Guilford, *Psychometric Methods* (New York: McGraw-Hill, 1954) (Chapter 7, "The Method of Pair Comparisons," and Chapter 8, "The Method of Rank Order"); H. Gulliksen, "Paired Comparisons and the Logic of Measurement," *Psychological Review*, III (1946), 199-213; J.B. Kruskal, "Multi-dimensional Scaling by Optimizing Goodness of Fit to a Non-Metric Hypothesis,"*Psychometrika*, XXIX (1964), 1-27; J.B. Kruskal, "Non-Metric Multi-dimensional Scaling: A Numerical Method," *Psychometrika*, XXIX (1964), 115-129; R.N. Shepard, "The Analysis of Proximities: Multi-dimensional Scaling with an Unknown Distance Function," *Psychometrika*, XXVII (1962), 125-139; L.L. Thurstone, "A Law

225

4. (contd.) of Comparative Judgment," *Psychological Review,*
XXXIV (1927), 273-286' W.S. Torgerson, *Theory and Method of
Scaling* (New York: John Wiley, 1958); and W.S. Torgerson,
"Multi-dimensional Scaling of Similarity," *Psychometrika,* XXX
(1965), 379-393.

5. L. Guttman, "An Approach for Quantifying Paired Comparisons and Rank Order," *Annals, Mathematical Statistics,*
XVII (1946), 144-163.

6. *ibid.,* 145.

7. J.B. Kruskal, *op. cit.*

8. This sample of 603 households was taken in the spring
of 1961 by the staff of the Statistical Service Division, Iowa
State University. The questionnaire was designed by Profes-
sors E. Thomas and W. Macki, and the survey was financially
supported by the Iowa College-Community Research Center and by
the Bureau of Business and Economic Research, University of
Iowa. The purpose of the study was to measure the economic
impact of the expenditure patterns of the rural population of
the state on towns of various sizes and at various distances
and to gain some insight into the probable effects of continued decrease in the rural population of the state on these
types of communities. A description of the sample with maps
showing the locations of the respondents can be found in
G. Rushton, *Spatial Pattern of Grocery Purchases by the Iowa
Rural Population* (Iowa City: University of Iowa, Bureau of
Business and Economic Research, Monograph No. 9, 1966),
Appendix A, 103-112.

9. The diagonal elements in Table 2 become greater than
zero whenever a locational type is chosen by a household which
has an alternative choice open to it belonging to the same
locational type.

10. Any empirical study contains some calls for which
there are no data, not because the sample size is too small,
but rather because in the area where the data were gathered
there may be several pairs of locational types which, whenever
they occured together, were always accompanied by a third
locational type preferred to by the other two.

11. W.S. Torgerson (1965), *op. cit.* 380, and, R.N.
Shepard, "Metric Stuctures in Ordinal Data," *Journal of Mathematical Psychology,* III (1966), 287-315.

12. This description of the scaling problem follows closely that of J.B. Kruskal (1964), *op. cit.;* and R.N. Shepard,
(1962), *op. cit.*

13. R.P. Abelson and J.W. Tukey, "Efficient Utilization
of Nonnumerical Information in Quantitative Analysis:
General Theory and the Case of Simple Order." *Annals, Mathematical Statistics,* XXXIV (1963), 1347-1369; and R.N.Shepard
(1966), *op. cit.*

14. M.G. Kendall, *Rank Correlation Methods,* 2nd Edition
(New York: Hafner Publishing Co., 1955), 156.

15. C.H. Coombs, *op. cit.,* 353-359.

16. *ibid.,* 347.

17. L. Curry, "Central Places in the Random Spatial
Economy." *Journal of Regional Science,* VII (Supplement, 1967);
R.G. Golledge, "Conceptualizing the Market Decision Process,"
Journal of Regional Science, VII (Supplement, 1967); R.G.
Golledge and L.A. Brown, "Search, Learning and the Market
Decision Process," *Geografiska Annaler,* XLIX, Ser. B. No.2.
(1967).

18 C.H. Coombs, *op. cit.,* 33.

19. C.G. Hempel, "A Logical Appraisal of Operationism,"
in *Aspects of Scientific Explanation* (New York: The Free
Press, 1965), Chapter 5.

20. H.S. Houthaker, "The Present State of Consumption
Theory: A Survey Article," *Econometrica,* XXIX, No. 4 (1961),
718. See also: W. Leontief, "Introduction to a Theory of
the Internal Structure of Functional Relationships,"
Econometrica, XV (1947).

21. A. Tversky, "A General Theory of Polynomial Conjoint
Measurement," *Journal of Mathematical Psychology,* IV (1967),
1-20; R.D. Luce and J. Tukey, "Simultaneous Conjoint Measurement: A New Type of Fundamental Measurement," *Journal of
Mathematical Psychology,* 1, (1964), 1-27.

22. W.S. Torgerson (1958), *op. cit.,* 59.

23. J.P. Guilford (1954), *op. cit.,* 192. The ratio N/n
indicates the average number of choices per stimulus. In
this study N/n was approximately 250.

24. R.N. Shepard, "On Subjectively Optimum Selection among Multi-attribute Alternatives," in M.W. Shelly and G.L. Bryan (eds.), *Human Judgments and Optimality* (New York: John Wiley, 1964), 257-281.

25. T.C. Koopmans, "On Flexibility of Future Preference," in M.W. Shelly and G.L. Bryan (eds.), *op. cit.*, 243.

THE MEASUREMENT OF MENTAL MAPS:
AN EXPERIMENTAL MODEL FOR STUDYING CONCEPTUAL SPACES

David Stea

Graduate School of Geography and Department of Psychology
Clark University

A. Introduction to Background Ideas

While Soviet psychologists have been directing attention
for some time to the mental representation of large environments, and to problems of location, orientation, and movement
within these environments, American psychologists -- strongly
rooted in a behaviorist tradition -- have shied away from the
same problems. Geographers interested in the emerging field
of "environmental perception," however, have become concerned
with the way people conceive of spaces varying in scale,
topography, usage, etc. Recently, attempts have been made to
"map" the characteristics of some of these mental spaces,
giving rise to the term "mental maps." The model that follows
is basically psychological, dealing with geographical material;
it draws heavily upon the theoretical concepts and constructs
of such psychologists as Kurt Lewin,[1] E.C. Tolman,[2] Jean
Piaget,[3] George Miller,[4] Andras Angyal,[5] and F.N. Shemyakin.[6]
The model attempts to establish a framework within which
measurement of the behavioral outputs of mental representations

may be given psychological and geographical meaning. It
treats spaces much larger than those usually dealt with in
American psychology, spaces so large that they cannot be apprehended ("perceived," in the orthodox psychological sense) at
once, nor in a brief series of glances.

The "model" which follows is experimental in two senses;
(1) it is a "first try" at reworking some older psychological
theory, and combining it with newer material, attempting to
account for observed conceptions of the physical world;
(2) it provides a "framework" for organizing a body of experimentation. At the moment, the "framework" aspect is more
important than the "model" aspect since only one part has been
modelled in a quantitative sense.

We make the following fundamental assumption: all persons
form conceptions of those significant environments that are
too large to be perceived, i.e. apprehended at once. These
conceptions may not themselves be spatial, but they somehow
order things which are spatial entities. Some of the ordering
principles are the following:

1. Establishment of hierarchies -- some places are more
important than others, for whatever reason;

2. "Boundedness": the space conceived ends somewhere;

3. Objects and places are located within the space, and
can be conceived as "points," more or less clearly defined;

4. The points exist in some relation to each other;
that is, one can speak of

    a. *distance* between points

    b. *bearing* from one point to another

    c. routes from one point to another, which include
the changes in direction taken in following the route;

5. Connectedness between points;

    a. a route existing between two points implies
connectedness;

     b. any interference with the route such as (1) lack of a transportation facility, or (2) formidable grades constitutes a barrier.

     The space of which we are speaking can be termed a "mental map," a "conceptual representation," or an "image" of the larger environment. Since it is a space too large to be perceived, the representation is a complex of things learned about the environment, including expectations, stereotypes, value judgments, etc.

     In physiological fact, we do not know that such maps exist. We have no reason to suppose that we will find patterns isomorphic to the larger world, or portions of the larger world, on the cerebral cortex. In the absence of further knowledge of how information is stored in the brain,[7] we have no direct test of the existence of such maps. The map model, or framework, is a construct: if the behavior of a subject talking about or depicting some aspect of the larger environment is described by the framework, then its utility is established.

     The experimental approach, then, takes two forms. In talking about orientation, and in talking of actual human movements, we are being behavioral. But, in our attempts to get at concepts of the larger environment by means of verbal or graphic responses, we are being basically introspective, and are bound to incur all the difficulties (and the criticisms) that such an approach entails.

     The new contribution of this paper is the suggestion that mental maps are measurable in a traditional sense, or, at least, that those behaviors (verbal responses, sketches, etc.) which operationally define the maps have discernible and interesting metric characteristics. We have set as our task that of measuring the deviation of a metric estimate from reality, or of one metric estimate from another.

Those characteristics of mental maps we hypothesize to be measurable include:

1. Absolute location of points
   a. distance, measured in (1) miles; (2) hours; (3) dollars
2. Relative locations of points
3. Extensity: size of an area and its shape
4. Strength of barriers

The eventual purposes of the research reported here, other studies currently underway, and contemplated future investigations, is to identify some of the independent variables which affect the formation of conceptual metrics. The metrics -- such as distance or extensity -- are viewed basically as dependent variables. The opposite view is also possible: conceived distance, for example, may affect estimates of accessibility.

Some of the independent variables hypothesized to affect conceptual metrics are:

a. The relative attractiveness of points viewed as origin and goal points, e.g., cities;
b. Noncommutative barriers;
c. Familiarity with certain trips;
d. Magnitude of the distance being estimated;
e. Kind and number of barriers separating end points;
f. Familiarity with certain areas.

Unlike some previous studies, those reported here do not aim at the production of entire maps, either verbally or graphically. Rather, the attempt is to elicit a component of a mental map with which the model deals, and to determine whether some hypothesized (but unmanipulated) independent variables relate to this component. The studies reported are not quite "minor" studies in the sense the term was used by psychologists in the earlier 20th century. Rather, they are pilot studies

designed solely to generate hypotheses which can be more rigorously tested in later designs.

The major tools used include verbal estimates of (1) distance, (2) time, and (3) direction; and graphic depictions
made by (1) drawing trip routes, (2) indicating points on blank
paper, and (3) indicating locations on outline maps.

B. Characteristics of the Model:

Preliminary Conceptualization

Cognitively organized large spaces[8] are hypothesized to
have the following "conceptual" characteristics (in terms of
their "mental representations"):

1. The behavioral output of the mental representations
can be elicited in the form of a space of "points" arranged in
some one-, or three-dimensional array. These points may possess dimension in actuality, and thus are not points in the
mathematical sense.

2. There may be several possible hierarchial arrangements among these, in terms of size, importance, desirability,
etc. In fact, several hierarchial arrangements may coexist
or may exist at different times within the life of an individual.

3. The space is somehow "bounded."[9] The boundaries may
be clear or indistinct: a neighborhood may be bounded by a
street or another neighborhood, for example.

4. If it is possible to get from one given point in the
imaged space to another by *some* means of transportation, they
are "connected," i.e., there exists a path between the points.

5. There may exist a "barrier" between any pair of
points.

  a. Barriers differ in their *permanence*, their
permeability, and their *quality* (e.g. they may be "natural" or
"artificial"). A construction project in the middle of a

superhighway decreases permeability, is impermanent, and "artificial"; reduces permeability to zero. A language change, while it may not *objectively* impede transportation, probably constitutes a conceptual barrier.

b. Barriers may be symmetrical (the same when approached from one member of a pair of points as from the other) or nonsymmetrical. The city of Providence, Rhode Island, for example, was a nonsymmetrical barrier to most travellers between Boston and New York prior to the extension of the turnpike and interstate highway system: that is, the north-south and south-north routes through the city were different and it was considerably more difficult to traverse the city in one direction than in the other.

In summary, the space of which we are speaking is bounded; one-, two-, or three-dimensional; and consists of a large though finite collection of "points" (of from zero to three dimension); of paths between them; and of interposed barriers.

We may also speak of "static" and "dynamic" maps. To form a "static" map, the subject need only imagine a space, and describe its "points," either in the form of free or directed recall, or by sketching a map from memory. We may then evaluate the relative salience of various points, when a number of subjects have been interviewed, in terms of verbal description or the physical placement of the points on a sketch map (both the relative positions of these points and their deviation from "reality" -- what might otherwise be termed "accuracy of memory"). This is the nature of the work reported by Kevin Lynch.[10]

The dynamic map differs from the static in that not only are the points of the map important, but so is the individual's *interaction* with these points -- how he imagines himself to be moving among them. We hypothesize that it is imagined or actual movement that ties the elements of the image together --

they may define the *nature*, if not the details of the image.
They may also answer some questions regarding human orientation
in a civilized world consisting largely of designed environments.

These "mental maps" need not necessarily be "geography in
the head"; in fact, as noted before, there may be no "record-
able" imagery at all without loss of heuristic value. It
matters not a whit that we cannot directly observe a "mental
map," or even that we cannot know for sure that it is actually
"there"; if a subject behaves *as if* such a map existed, it is
sufficient justification for the model.

What is proposed is that the large and complex real world
must be handled by people with limited capacity for information
storage, manipulation and retrieval. Hence, the individual
makes certain simplifications and adjustments, in accordance
with his needs and experience in the conceptualization of these
large and complex spaces. His adjustments are distortions
only in the sense that they produce a conceptual space that
does not correspond directly with geographical maps. The
magnitude and direction of distortion may be expected to vary
from individual to individual, but those distortions of most
interest will be the ones consistent across a given class of
persons. The measures used to determine simplification or
adjustment may be behavioral, in the sense that we observe the
individual's actions *vis-a-vis* a real or artificially created
environment and infer the "image" that yielded the observed
behavior, or "introspectively descriptive" in that we ask him
to describe an *imagined* interaction with the environment.[11]

235

C. Three Elements of the Preliminary Model:

Distances, Bearings, and Turns

We begin by asking the most general question, "What does a person do in conceptualizing a space?" First, he is or imagines himself in a location, that is, at a "point" (origin) in space. Second, he moves, or imagines movement to another "point" (goal), probably passing through intermediate "points" (which may be termed sub-goals) on the way. These various points bear a certain relation to each other, the weakest of which (in correspondence with the real world) is the preservation of order -- which point comes first, which second, etc. in the direction of travel. Third, the origin and goal are separated -- there exists some distance between them. Fourth, movement is initiated in a certain direction (bearing) which is presumably related to the imaged location of the goal *vis-a-vis* the origin. Fifth, this bearing is generally not maintained throughout the entire journey or "virtual trip"; changes in direction which we call turns periodically occur; and, in order to perform these changes in direction, the person usually operates on the basis of some system of coordinates.

Distance

The conceptualization of "distance"[12] in spatial imagery may be aided by phrasing the issue in quasi-mathematical terms. The "real world" is, mathematically speaking, a metric space, that is, if we take any non-trivial portion of it and call it "X" ("X" may be a (non-mathematical) neighborhood, a region or part of a nation), we can proceed to consider a function defined on the non-empty set "X". This function will be called "d" (distance), and the pair (X, d) is then a metric space if the following three conditions are satisfied for any pair of points (a and b) in "X" (for d (a,b) read "the distance between "a" and "b").

(1) $d(a,b) \geq 0$; $d(a,b) = 0$ if and only if a = b.

(2) $d(a,b) = d(b,a)$.

(3) $d(a,c) \leq d(a,b)$ $d(b,c)$: the familiar "triangle inequality."

An interesting question is whether the image world is metric in the above sense. In the next section, we present some preliminary evidence obtained by Irene Buckman[13] from a recent study of "virtual trips" among cities in Southern New England in which a person imagines that he is travelling by a specified means of transportation between two points known to him, and indicating that condition (2) does not necessarily hold in imagery; it makes a difference whether the question asked is: "What is the distance between Providence and Boston," or "What is the distance between Boston and Providence." That is, distance estimates appear *noncommutative*. While there is no available empirical evidence on (1) and (3), it might be interesting to speculate on the meaning of failure of these conditions to hold. Several ideas have been suggested, but space limitations prevent detailed explication here.[14]

Bearings

In addition to impressions of distance the mental map must include an impression of the initial direction taken from the point of origin or the position of the person at the origin, to the goal. The "perceived" straight line or geodesic connecting one "fixed" point with a second and the "perceived" bearing of the second from the first uniquely define the representation of their relation. We define two kinds of bearings in accordance with two hypothesized modes of orientation:

(1) Body or "ego-centered" bearing -- directions given in terms of the individual's position, i.e., left-right, back-front, up-down, etc.

(2) "Objective" bearing, which can be further divided into "Universal," "Macrolocal" and "Microlocal" compass direction.

Universal direction is based upon compass or map coordinates.
However, there is a distinction between universal and consensual
north -- that is, what is considered to be north often varies
with locality.[15] The locality may be as large as a region, in
which case we term the consensual coordinate system "macrolocal," or as small as a town, city, or metropolitan area, in
which case we speak of "microlocal" compass direction.

Turns

A turn is simply a change in bearing and may be described
with reference to a previous bearing or direction of travel
(e.g., "turn right") or with regard to compass coordinates.
But the most interesting characteristic of an "imaged" turn is
its magnitude. With an infinity of variations possible,
people tend to simplify or "adjust" those imaged turns. We
hypothesize that the most common adjustment of a turn for
dwellers in gridiron cities is a right angle; for dwellers in
those cities in which a radial pattern is superimposed upon a
grid (or vice-versa), as is the case with Washington, D.C.,
adjustments of 45° and of 90° may be made. Thus, turns may
well be remembered as other than what they are, and the nature
of simplifying adjustments is likely to be a result of one's
accustomed interaction with the environment. Thus, we hypothesize that people whose experience has been largely with cities
characterized by "irregular" lines of communication -- such
as Guanajuato in the Republic of Mexico -- are likely to have
received the least reinforcement for "simplifying assumptions."
Further, they will probably experience much less difficulty
than those persons who are accustomed to more regular patterns of paths, when the pattern of lines of communication
departs slightly from rectilinearity. A psychophysical question
may be posed in this context: for a given individual or group
of individuals, how large does a departure from strict

rectangularity have to be before it is perceived (treated as) a departure?

D. Initial Work on Problems Related to the Model

Our research has thus far been confined to five "minor" studies -- minor in the sense that they have been only preliminary investigations of some of the model's predictions. Each of these has involved a paper-and-pencil inventory of one sort or another, with all the attendant drawbacks, rather than direct measures of behavior. The primary phenomena investigated have been distance estimates (both in traditional measures of distance and in time), compass direction estimates, relative sizes of geographical areas, and location. The subjects of these experiments were University students in Providence, Rhode Island, Worcester, Massachusetts, and San Cristobal Las Casas, Chiapas, Mexico. Although no tests have yet been made of the statistical significance of the results of these studies, they yield some indication of support for predictions derived directly or indirectly from the model:

(1) In general, distances are noncommutative, the non-commutativity depending upon

      (a) the relative valence of the goal and starting point

      (b) the "barrier" aspect of the connecting path

      (c) the "attractiveness" of the connecting path

      (d) the subject's familiarity with the trip and its end points

(2) In the estimation of geographical areas

      (a) if a person is totally foreign to all the areas, he tends to estimate accurately those about which he has most information

(b)   in the estimation of the relative sizes of
familiar areas, the most familiar and most
valued are also those most likely to be overestimated.

These findings, and others, are described in detail below.
It should be borne in mind that the results of paper-and-pencil
inventories are valid as predictors of behavior only insofar
as test behavior relates to environmental behavior.

Short distances -- New England and Chiapas

A study was made at Brown University by Irene Buckman[16]
to test the hypothesis that under specifiable circumstances
the conception of distance from a less preferred to a more preferred city is smaller than the conception of the same distance
in the opposite direction. Forty male and female students
were asked to rank six New England cities and New York in order
of preference (numerical rankings were used rather than paired
comparisons to avoid involvement in problems of transitivity).
All twenty-one possible combinations of city pairs were then
formed, and two lists made: the first consisted of one set of
twenty-one permutations, the other of the complementary set.
Thus, for every city pair in one direction on one list, a pair
was specified in the opposite direction on the other list.
Twenty subjects were asked to estimate the distance and compass bearing between pairs of cities in one direction of travel,
and twenty subjects were asked to make similar estimates in the
opposite direction of travel. Each subject was also asked
(1) how many times he had visited each city, (2) how many times
he had taken each trip by road, and (3) how many times by air.
Only two of the forty subjects owned cars.

Each of the forty-two possible trips, then, could be in a
preferred or non-preferred direction for a given subject (preferred direction is here defined as having the preferred city
as a goal). For five of the forty-two trips, one city of the

pair was preferred by all twenty subjects in the group: hence, people who preferred one terminal city could not be compared with people who preferred the other. Two trips yielded equal distance estimates for the two directions of preference; one more yielded questionable data. Of the thirty-four trips which met the criteria of (1) having subjects in both preference groups and (2) yielding differing mean estimates of road distance, both differences between the estimated and actual road distance and the ratios of the differences to the actual distances were computed. Of these thirty-four trips, twenty, or 59% yielded shorter estimates for trips in the direction of the preferred city -- not a terribly impressive result, treating the data as nominal. However, for those subjects who estimated the trip in the preferred direction (to the preferred terminal point) as longer than the trip in the opposite direction, the mean magnitude of error (estimated-actual distance) was 15.5 miles, while for those who estimated the trip was shorter, the mean magnitude of error was over fifty-four miles. In this case, ratios are probably better indicators of error, and have yet to be compared.

A second interesting result emerged from the above. Subjects for whom trips were relatively familiar (those groups for whom the average number of trips actually taken between the two points was one or more) were compared with those who averaged less than one actual trip. It appears that the more familiar a subject is with a given trip, the more accurate is his estimate, *but* the more discrepant are his estimates in the two directions of travel. This is in line with the predictions of the model, in which we are forced to assume that increasing acquaintance with the two end-points of a trip will increase the difference in their relative attractiveness.

In measuring the commutativity of estimated distance, it would be ideal if we could make same-trip comparisons within a single subject. Unfortunately, people have the disconcerting

characteristic of remembering the estimates they have made.
It might be possible, however, to administer tests to subjects
at one time asking for estimates in one direction, then to ask
for estimates in the contrary direction at another time --
assuming, of course, that the subjects do not engage in geographic self-education in the interim as a result of the first
test experience.

Hansen's Chiapas study[17] utilized three cities -- Tuxtla
Gutierrez, San Cristobal Las Casas, and Comitan, connected with
each other only via the Pan-American highway. All of the 156
subjects in the study were college students studying in San
Cristobal Las Casas. The road distance from Tuxtla, in the
west, to San Cristobal is precisely identical to that from
San Cristobal to Comitan, in the east, (eighty-five kilometers),
but the routes are markedly different: the road from Tuxtla is
tortuous and climbs almost 8,000 feet before dropping down again
to San Cristobal, passing through several climatic zones; the
road from Comitan is much less winding -- a good portion of it,
in fact, is completely straight -- and rises much less in alti-
tude. In the New England study, we assumed that the "barrier"
effect of routes was relatively commutative, that the negative
effect of having to make a trip was equivalent in both trip
directions, and that the major contributing factor to the non-
commutativity of distance estimation would be the difference in
attractiveness of the two end points. In the Chiapas study,
it was assumed that the major contributing factor to differences in distance estimation -- Tuxtla-San Cristobal vs.
Comitan-San Cristobal would be the different "barrier" effects
of the two roads, the one from Tuxtla to San Cristobal being,
objectively, much more difficult to traverse. Our prediction
was that the distance easier to traverse would be perceived as
the shorter. The result was contrary to expectations, for
fully two-thirds of the sixty-six subjects who utilized the

same primary and secondary modes of travel (public transportation) on trips among the three cities estimated the distance
from Comitan to San Cristobal as greater than that from Tuxtla
to San Cristobal. It was this apparently paradoxical finding
that led to the insertion of 1(c) as a *post-hoc* "prediction":
anecdotal reports indicate that the trip from Tuxtla to San
Cristobal is more beautiful, more interesting, and divisible
into more "cognitive segments" than is the trip from San
Cristobal to Comitan. Some evidence in the psychological
literature suggests that more interesting, more "divisible"
experiences are perceived as of shorter duration.[18] Since
these students travelled generally by bus, they did not directly experience the arduous effort of driving. An interview
of bus drivers might have elicited markedly different results.

Long distances -- global estimates

A group of twenty-two Clark University graduate and under-
graduate students taking a course in Behavior and Environment
estimated the great circle air distance and travel time between
New York and eleven other cities of the world.[19] Half of these
were presented with the problem of estimating the distance and
time "from New York to (City "X")" and the other half with the
problem of estimating the distance "from (City "X") to New
York." The results indicate the following:

(1) Considering the Eastern and Western hemispheres, distances and times between cities within a hemisphere or distances
objectively less than 4,000 miles tend to be overestimated;
distances between hemispheres or objectively greater than 4,000
miles tend to be underestimated.

(2) There exists the possibility that "the way home" is
perceived shorter in distance, and especially in time, than the
trip going away (most of the subjects' homes were within a 200-
mile radius of New York City). No conclusions are possible,
however, because a given subject made *all* estimates either

going away from or toward New York, and the two groups were not equivalent in the overall accuracy of their estimates, in terms either of distances or time.

(3) Classifying barriers between cities into three groups: geographical (if the cities are not connected by land), political (located in independent nations), and linguistic (speaking different languages):

    a. Six pairs of cities were separated *at least* by geographical, eight by political and five by linguistic barriers -- there was no difference whatever among distance (and time) estimates for these three groups of city pairs;

    b. When the *number* of barriers separating members of a pair was considered, there was a suggestion in the data that the distance (and time) between city pairs separated by no barriers or by one barrier tends to be underestimated, while that between city pairs separated by two or all three barriers is overestimated relative to actual distance. Since the mean distance between cities separated by no barriers or one barrier was 3,186 miles and that between cities separated by two or more barriers was 4,047 miles, it seems that the "barrier effect," if there is one, works in opposition to the simple "distance effect" suggested in (1) above.

The above results then, provide hypotheses to be more rigorously tested in future experimentation.

Estimations of areas -- local scale (New England)

D. W. Griffin[20] suggests that not only are distances and directions distorted in what he terms people's "topographical schemata," but relative areas of various regions, too, reflect the importance of these areas to the individual rather than geographical reality. This suggestion, which forms the basis of the predictions previously listed under "2," was the impetus for a study of conceptual maps of New England.[21] Twenty-three geography students (twenty males and three females) and twenty-three nongeography students (seventeen male and six female) were instructed to indicate those locations *they thought they knew* of six specified points in each of six New England states on a sheet of paper. The paper was blank except for a title ("New England States"), a scale, Magnetic North, and a point indicating the location of Worcester, Massachusetts. Subjects were asked to draw an arrow to each of the points he has located, and to estimate the air distances to those points.

The overall result was a "vector field." The mean compass direction and distance estimate for each location was computed separately for geographers and nongeographers, and these were projected onto an outline map of New England -- the maps previously removed from the paper given to subjects to draw upon. "State boundaries were judged by assigning proportionate distances from those points roughly defining the outline of the States (deliberately chosen for this property and for being relatively well known and identifiable with their respective States) to the border or coastline itself. In this manner the conceptualized points were amalgamated to form conceptual (cognitive) maps for each of the two groups." These "eyeballed" results (visually estimated with the aid of an Atlas) are depicted in Figures 1 and 2. The two groups did not differ markedly either in mean age or in mean duration of residence in New England, although the nongeographers were slightly older and had resided slightly longer in New England than had the geography students; this may explain in part why the amalgamated nongeographer's map appears to approximate physical reality more closely.

In light of our earlier prediction -- that among familiar areas, people tend to give exaggerated estimations of areas most important to them -- it should be noted that areas of Connecticut and Massachusetts appear to be overestimated by both groups, relative to the estimated sizes of the other New England States. Further, "in the nongeographer's maps (Fig.1)

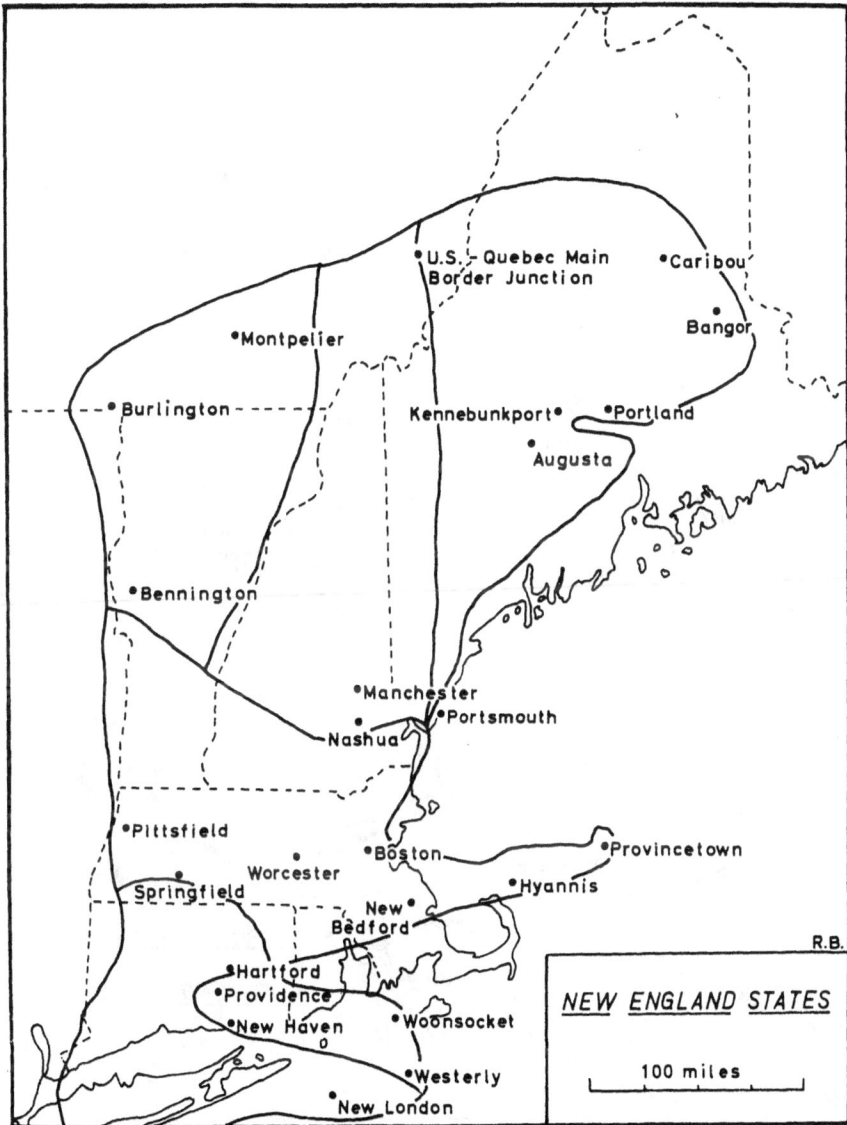

Fig. 1   Non-geography students'
collective conceptualization

246

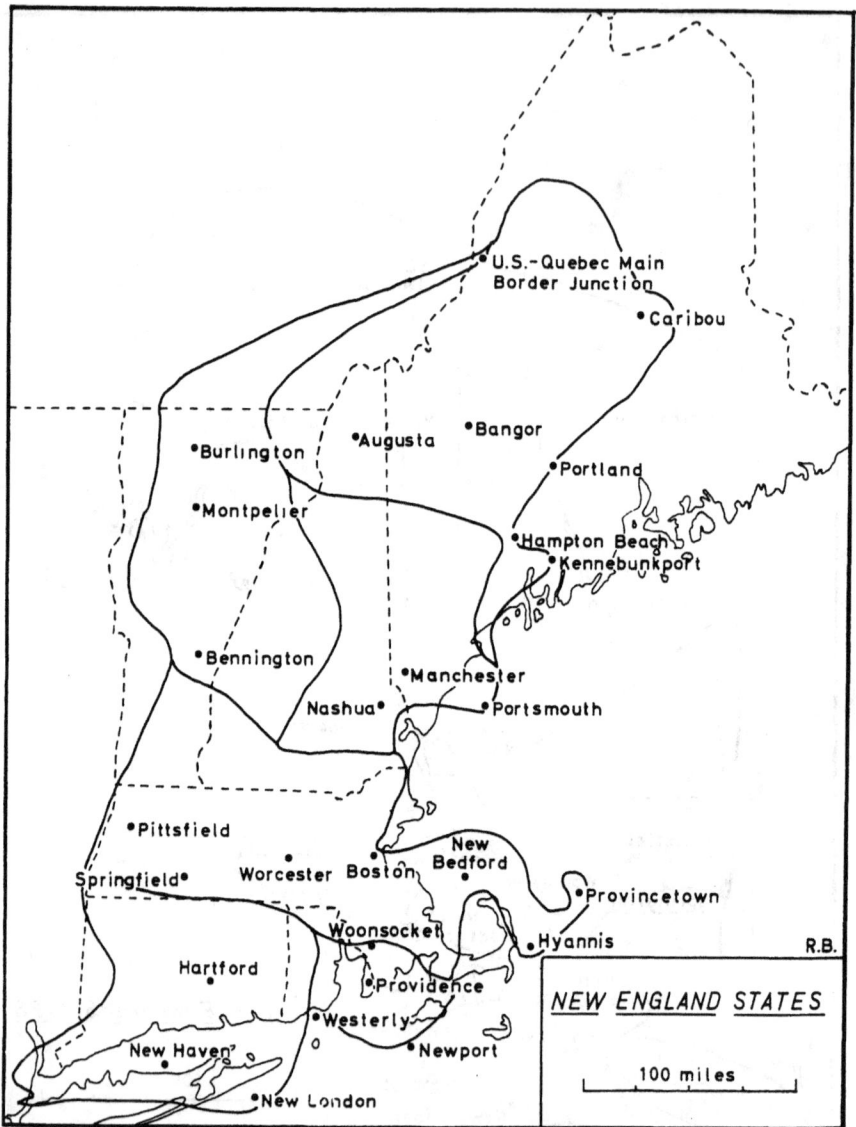

Fig. 2 Geography students'
collective conceptualization

... Connecticut is...roughly twice its actual proportionate
size; most of the nongeographer's home-town residences are in
Connecticut."

Estimations of areas -- global scale

Twenty Clark University students, eleven females and nine
males (fourteen of whom had no formal training in geography on
the college level) were asked, among seven questions, to estimate the shortest distance from Worcester to each of ten for-
eign countries:[22]

Brazil	Viet Nam
Israel	South Africa
Greenland	Egypt
New Zealand	Japan
Cuba	Sweden

They were also asked to estimate the relative sizes of the
countries, both in ordinal fashion and in comparison with the
size of New England. It was hypothesized that "countries which
are 'trouble spots' (Viet Nam, Israel, Cuba, etc.) would be
more accurately estimated than neutral countries (New Zealand,
Sweden, Greenland)."

While distance estimates on this global scale were found
to be unrelated to judged importance or objective distance,
perceived *size* and judged economic and political importance
appeared to be highly correlated -- more so, in fact, than
perceived and actual size. "The first four countries in importance (Viet Nam, Japan, Israel, Cuba) rank in the top four
in correct size perception. Excluding Brazil, the next three
important countries (Egypt, South Africa, Sweden) rank next
in a bracket for correct size perception and New Zealand and
Greenland can be bracketed on the end."

It is tempting to say that there exists a relation between
these findings and results of psychologists' studies of the
influence of motivation upon perception.[23] Too many potentially

248

confounding factors exist, however. In view of another suggestion of the above study -- that experience with the Mercator
Projection of the globe in fact exerts a strong influence upon
estimates of direction -- we may assume that its influence extends to size and distance estimates, as geographers have
hypothesized in the past. If this is true, then it is important to note that the countries whose size was most accurately
estimated all lie about the region of least distortion on the
Mercator. They are all "small" countries, as well, generating
the additional hypothesis that the extents of small land areas
are more accurately estimated than large ones, if both are
relatively unfamiliar. Finally, there exists a question of
relative familiarity: we cannot say that perceived importance
causes accuracy of size estimation, since both are clearly influenced by familiarity (aided or indexed by the extent to
which information, often graphic, is conveyed by the mass
media).

E. Toward a Taxonomy of Schemata

-- Developmental Implications

The notion of a "spatial schema" has been written about
by Griffin[24] and Lee.[25] Lee states:

...we assemble models in our heads that are constantly
being modified by new and relevant experience, but
which exist as entities...with their own synamic organization....*schemata are related to, but by no means
coincident with, the physical reality that lies outside us.* (A) problem is how the subjective schemata
develop from the dimensions of "real" space. Such
studies that have been done to trace the growth of
spatial representation confirm that it is not something that is immediately given -- it has to grow gradually and meticulously with the child, and apparently
as a result of his experience with objects, seeing them,
handling them, observing their movements. Piaget thinks
that adults build a grid of coordinates in their heads
to orient themselves in space. When a number of school

children in Devon were asked to estimate the distances from their homes to the major cities, and especially to London, it appeared that their spatial
world was divided into various local schemata which
bore at least a detectable relationship to the physical world, but that outside this there was one total
schema that might be called the "elsewhere" scheme,
in which physical dimensions were irrelevant. A
study in the United States showed the subject children to have a local "direct experience" schema, and
beyond this they had, if old enough, a knowledge of
a conventional set of mileage numbers which they had
begun to associate with names of cities. This, like
a good deal of geography as it is taught at present,
is rote learning and not spatial cognition.[26]

This suggests that the development aspects of schemata
ought to comprise an important field of study, not restricted
to the way in which children build up concepts of conceptual
spaces but including the manner in which adults who have recently arrived in an area build up representations of that area.

F. Conclusions, Qualifications, and Suggestions for Future Research

Further studies might profitably deal with scales smaller
than those treated here. As a problem area, the city is of more
interest today than ever before. Can urban imagery be quantified? Is there variation in the perceived extensity of a sub-
area of a city associated with ethnic group or socio-economic
group identification with the area? How variable are the
"conceptual boundaries" of an area? Studies we have performed
in Mexican cities indicate that the variation is tremendous.
Is the willingness of a given person to travel to a given job
location associated with measureable conceptual distance?

On an even smaller scale, the "architectural scale" in
designer's terminology, vertical distances and discrepancies
among them assume importance in proportion to relatively short
horizontal distances. How does "up" distance compare with

"down" distance? Is it possible, using conceptual, non-physiological, measures to quantify the effort variable differentiating ascent from descent under a variety of design conditions? Are there differences between the perceived extensities of "outside" and "inside" distances, i.e., does the process of exiting from one part of a building and entering another part appreciably alter intra-building distance estimates? What is the effect of seasons and associated weather conditions upon such estimates?

Since it has neither been axiomatized nor formalized symbolically, the framework presented here is a model in the conceptual sense only. The model specified certain measurements that we ought to be able to make on mental maps, but is mathematical only in the case of conceptual distance. Postulated relationships among bearings and among turns have yet to be quantified.

The assumption implicit in the model presented here is that people behave in accordance with spatial schemata -- mental representations -- of "physical reality," and that knowledge and understanding of these representations should be of importance to geographers, architects, and planners. Clearly, these representations develop as the organism experiences and matures; we have yet to determine just how. This discussion provides some suggestions. Current research is sharpening our hypotheses; future research will increase the model's sophistication, and even further effort may result in its admission to the world of theory.

251

NOTES

1. K. Lewin, *Field Theory in Social Science* (New York: Harper, 1951).

2. E. C. Tolman, "The Model." Chapter 2 in Part III of T. Parsons, and E. A. Shils (eds.), *Toward a General Theory of Action* (Cambridge: Harvard University Press, 1951).

3. J. Piaget, and B. Inhelder, *A Child's Conception of Space* (New York: Norton, 1967).

4. G. A. Miller, E. Golanter, and K. H. Primbrom, *Plans and the Structure of Behavior* (New York: Henry Holt, 1960).

5. A. Angyal (unpublished lecture at Massachusetts Institute of Technology, 1965).

6. F. N. Shemyakin, "Orientation in Space," *Psychological Science in the U.S.S.R.*, Vol. 1 (Washington: Office of Technical Services, 62-11083, 1962).

7. See J. Hockberg, *Perception* (Englewood Cliffs, N. J.: Prentice-Hall, 1964) for further comments.

8. The term "mental map" is used in a somewhat different sense by Peter Gould: this usage is described in "On Mental Maps," Discussion paper, Michigan Inter-University Community of Mathematical Geographers (multilith); and in other works by the same author, such as: "Structuring Information on Spacio-temporal Preferences," *Journal of Regional Science*, 7 (2) (Supplement, 1967).

9. The classic geographical work on boundaries is represented by J. R. V. Prescott's *The Geography of Frontiers and Boundaries* (Chicago: Aldine, 1965). Recent commentaries more pertinent to this discussion include John Nyestuen, "Boundary Shapes and Boundary Problems," *Peace Research Society (International) Papers* 7 (1967), and Robert Yuill, "A Simulation Study of Barrier Effects in Spatial Diffusion Processes," Discussion paper, Michigan Inter-University Community of Mathematical Geographers (multilith).

10.   K. Lynch, *The Image of the City* (Cambridge: M.I.T. Press, 1960).

11.   To be termed the "virtual trip".

12.   "Distance" in its most general usage implies only a separation between points: the metric may be miles, hours, dollars, or other units of effort involved in getting from one to the other.

13.   I. Buckman, "The Metrics of Psychological Space: An Experiment" (unpublished ms., Brown University, 1966).   For a discussion of metric spaces see any text discussing topology, e.g., Spanier's *Set Theory and Metric Spaces*, University of Chicago (mimeo.)

14.   A more general study of spatial distortion was recently completed by Ralph Lennon, using students of Quinsigamond College as subjects: R. Lennon (unpublished paper delivered at Clark University, 1967).

15.   In 1913, Trowbridge, reprinted in I.P. Howard, and W.B. Templeton, *Human Spatial Orientation* (New York: Wiley, 1966), asked subjects to indicate the direction of geographical locations relative to the center of a sheet of blank paper; he dichotomized his findings into the conventionally-oriented (those who based their judgements on "objectives" -- e.g., compass -- coordinate systems) and the egocentrically oriented (those whose referents or coordinate systems were subjective -- e.g., body centered).   Angyal, *op. cit.*, has noted a similar dichotomy.

16.   Buckman, *op. cit.*

17.   I.D. Hansen, "Perception of Travelled Distance" (unpublished ms., Clark University, 1968).

18.   However, there is other evidence to the contrary. Terence Lee reports a study in which were presented "on large pieces of white card, a number of lines, of one of three lengths, 6, 8 or 10 inches, but with from one to five right-angled corners along their length.   The subjects were asked... to reproduce their judgements of the length of the lines by drawing a single equivalent but straight line...this experiment showed a clear and consistent overestimation in the judged length as a function of the number of corners."   See T. Lee, "Psychology and Living Space," *Transactions of the Bartlett Society*, 3 (1966), 11-36.

19. D. Stea, "Estimation of Distances on a Global Scale" (unpublished ms., Clark University, 1968).

20. D. Griffin, "Topographical Orientation." In E.G. Boring, H.S. Langfeld, and H.P. Weld, *Foundations of Psychology* (New York: Wiley, 1948).

21. J. Dowd, and D. Zaido, "Conceptual Maps of New England" (unpublished ms., Clark University, 1968).

22. C. Eaton, and K. Lawrence, "People"s Macroimagery of the Environment (Images of the World)" (unpublished ms., Clark University, 1968).

23. J.S. Bruner, and C.C. Goodman, "Value and Need as Organized Factors in Perception," *Journal of Abnormal and Social Psychology,* 42 (1947), 33-44.

24. Boring, Langfeld, and Weld, *op. cit.*

25. Lee, *op. cit.*

26. Lee, *op. cit.,* 23-26.

PROBLEMS IN MODELLING INTERACTION:

THE CASE OF HOSPITAL CARE

Richard L. Morrill

University of Washington

and

Robert J. Earickson

University of Hawaii

The student of location problems, accustomed to the com-
plexities, say, of industrial location or multi-purpose shopping trips, might naively imagine that trips to hospitals
would at least offer a simpler vehicle for tests of theories
of spatial behavior. Unfortunately, when the real world of
hospital use is examined, one is faced with a pattern of
great complexity, apparent confusion and irrationality.

As with any behavior problem, however, much underlying
order can be found.   In the case of patient use of physicians
and hospitals, very simple models incorporating optimizing
principles of spatial, social and economic behavior are able
to account for much of the variation in behavior, and can produce the pattern of use to a degree.   But they do fall short
of a fully satisfactory understanding of the processes and
patterns.   In our study,[1] therefore, we were persuaded to use

the simulation approach, because experimental methods were needed in which hypothetical modes of behavior could be varied in an attempt to discern which better characterized actual decision-making. The resultant simulation model may be operated either deterministically or probabalistically -- the latter in order to allow for a degree of irrationality or uncertainty, and also to handle the problem of indeterminacy or patients and physicians being confronted with decision between approximately equally good choices.

If we are studying large aggregates of patients, physicians and hospitals, we can employ simple economic models to evaluate needs for beds or personnel.[2] Suppose, however, we discover there is a shortage of capacity in the system. We are now forced to disaggregate to the individual hospital and small segments of population in order to determine where capacity should be added. It is interesting to observe that, in general, distance minimizing behavior is supported by aggregate patterns of patients around hospitals,[3] but the real world of planning requires the use and viability of individual units. Inevitably, then, more interesting problems are at the micro-level and simulation approaches are appropriate.

Specifically, we find that gravity, interactance-type models may be employed for a general description of the use of the hospital system, and to a degree to indicate imbalances in the location of hospital capacity.[4] Distance-minimizing transportation models fail to allow sufficient flexibility in behavior, but they also prove valuable in identifying groups of patients which are poorly served or hospitals which are poorly located. The simulation models were devised to allow for real variability in behavior and substitution among alternatives while retaining the ability to evaluate the efficiency of the system. Simulation models can identify behavioral processes consistent with the production of particular spatial

patterns.[5] But, their value is greatly enhanced if we can
extend the model to test the effects of possible relaxations
in the decision controls, and to estimate shifts in location
of capacity that would raise the level of satisfaction of
patients, physicians, and hospitals. By satisfaction we argue
that 1) patients desire easier access in space and time to
physicians and hospitals with desired characteristics;
2) physicians desire both good access to hospitals and full
use of their capacities; and 3) hospitals desire a high rate
of occupancy without congestion and excessive waiting. Any
evaluation of the hospital system needs to be able to identify
groups of patients and physicians who are required to travel
unusually far, or suffer unusually long waits due to congestion. It should be able to identify hospitals which have
excessive or deficient demand for their capacity, and to pre-
scribe shifts in physician and hospital capacity that would
bring all patients within minimum travel times -- to be
specified by society ultimately -- and also all hospitals to a
viable level of operation.

A. Nature of the Hospital Use Problem

Our procedure was to analyze the actual use of the hospital
system in order to estimate how it should be improved. Ideally,
the problem could be reduced to one of sending patients from
their homes to hospitals according to notions of patient evaluation of the hospital's attractiveness, and some function of
size and distance.

The first complication is that for most patients the trip
is a two-stage one, not necessarily at the same time. The
patient visits a physician, who, in turn assigns the patient
to a hospital.[6] While we know that patients have opinions
about some hospitals, we also know that, overwhelmingly, the
physician, not the patient, decides the hospital. Commonly,

the decision benefits the physician more than the patient; that is, the hospital is apt to be closer to the physician than to the patient. In any event, while the patient trip to physician tends to reflect patient evaluation of the number and distance of physicians, the hospital visit will tend to reflect both the patient and physician evaluation of the distance and size of possible hospitals. They physician role here obviously helps explain why many patients travel beyond hospitals closer to them.

In addition, some patients cannot afford to go to physicians at all, nor can they afford to pay for hospital care. To the extent that they nevertheless must have care, their needs are met by only a small portion of possible hospitals -- a few of which are designed only for charity patients, and a few of which will take some charity patients, as for medical interest, under state or federal subsidy.[7] Such a restriction tends to force charity patients to travel further than they would otherwise need to. The present model will be able to evaluate the improvements from relaxation of these restraints.

Again, patients differ in the severity of the medical problem and hospitals differ in their ability to handle more esoteric demands.[8] In a large metropolis, at least four kinds of hospitals are discernible. A few very large teaching and research hospitals which can treat the most unusual cases; a few large hospitals which can treat a variety of intermediate level problems; many smaller hospitals which are equipped to handle the normal range of problems, and a few special-purpose hospitals, that will handle but one kind of problem. The effect is again to restrict choice, especially for the patient needing more specialized treatment. Only the largest and most centrally accessible hospitals can afford the high costs of providing a higher level of care. Thus, many of the trips to such hospitals from seemingly irrational distances are reasonable and perhaps inevitable.

Besides level of care, some distinctions on type of care
are fundamental. In general, children require separate units
in a hospital, as do obstetrical patients. Not all hospitals
have wards for these classes of patients, while a few hospitals
may emphasize them. Choice for some patients is thus restricted
and here little or no substitution is legally or medically possible. On the other hand, pediatric and obstetric facilities
often tend to concentrate in areas of younger population.

The fact of race, notably black and white, also has dra-
matic influence. The dependent Negro population is restricted
to the few hospitals taking charity patients, but even the paying patient will tend to be treated by Negro physicians who in
turn are likely to be restricted to a few all-Negro hospitals.[9]
Clearly, the Negro patient does not have equal access to hospitals and must travel further on the average. Some limited substitution is possible such as between the close all-Negro hospital and distant hospitals which will at least accept them.
The model will be able to evaluate the improvement from possible elimination of this discriminatory restraint.

Finally, religion is found to play an interesting role, pre-
senting patients with such choices as between a distant hospital operated by the same religion as his and close hospitals
of different persuasion and hence less preferred. In the Chicago area at least, many patients clearly were willing to
travel somewhat further to reach Protestant, Catholic or Jewish
operated hospitals.[10] Thus to the extent that regligion is
important to people in the seeking of physician and hospital
care, we need to evaluate the adequacy of hospital care for
the population which so discriminates.

To summarize: we need to recognize that patients require
different levels of care, different kinds of care (pediatric,
obstetric, general medical-surgical), and that hospitals will
be similarly differentiated; that patients will differ in race

and in ability to pay; and hospitals in their willingness to care for charity or Negro patients; and that patient differences in religion lead some to perceive advantages in hospitals under control of their denomination.

The problem of modelling hospital use and evaluating needs is thus partly met by disaggregation of patients and hospitals by level of care, type of care, ability to pay and race. For these, only a very limited amount of substitution is possible. But religion does not pose such an absolute barrier -- the Catholic population does not have to go to Catholic hospitals. The behavior of patients (or physicians) is not strictly distance minimizing, but depends on the willingness to substitute greater distance for larger size and presumed higher quality. Thus it is primarily because of religious variation and indeterminacy about what is the 'best' destination that requires us in the end to use the simulation approach.

B. Utility and Limitation of Simple Deterministic Models

Distance-Minimizing Programming Models

One simple hypothesis of spatial behavior is embodied in the 'transportation problem' -- here that patients go to the nearest available physician (or hospital) which competition permits. Distances travelled by all patients to all physicians -- that is, system distance -- is minimized. Discipline is enforced by the dual-produced structures of prices.[11] For example, poorly located hospitals with much capacity are called upon in effect to offer lower prices in order to attract sufficient patients, or the small hospital in a growing suburb would need to or could raise prices to discourage overcrowding. Some small evidence for such price variation exists. There seem to be some longer movements to hospitals

with lower occupancy in order to avoid waiting time at over-
crowded hospitals. But in general, prices reflect internal
hospital costs, not competition for patients.[12] It is also
unlikely that patients evaluate possible destinations according
to the interests of the system, so the model is really not the
most appropriate.

The main problem, as those who have worked with such models
might guess, is the severe simplicity of solution (a mathematical necessity). Very few routes are permitted; indeed
all the patients of most areas will be sent to but one destination, even though other destinations may only be marginally
less satisfying or even equally good (that is the solution may
well be partly indeterminate). Patients are allowed no substitution; all must behave alike. Thus as a reproducer of
actual flows, the transportation model is hopelessly too simplifying. Certainly more and more paths will be generated
the finer the disaggregation of groups and hospitals. Nevertheless, route limitation will always occur. In comparing
predicted and actual flows of Negro patients to Negro hospitals,
for example, the model could only predict about one-third.[13]
Most of the other paths to hospitals actually used were not
irrational but only slightly farther.

It is possible to modify the transportation model to permit more paths. The optimal solution may be considered to
apply to some larger proportion of the patients; and then
second and perhaps third, etc., suboptimal solutions allocate
remaining patients, with however, the originally used paths
made more difficult to use. Unfortunately the procedure is
somewhat arbitrary, and there remains doubt about whether
people perceive choice in this way. We are continuing to work
on such modifications but without high hope. One escape is
via the imposition of maximum permitted flows on particular
routes, but that amounts to sheer contrivancy.

Despite these descriptive shortcomings, the transportation
model still has prescriptive utility. An optimum solution,
for example, a patient to physician 'run' clearly depicts which
sets of patients have to travel farthest for care, and which
physician clusters have to depend on distant patients. The
'shadow price' structure reflects the pattern of flows in a
simplified way, and constitutes evidence that if we reduce
capacity in the 'worst off' physician clusters and add capacity in physician clusters nearest the 'worst off' patients,
we will be able to reduce the time and cost of travel. In one
experiment, for example, we discovered that a shift of only
15% of physicians (mainly to growing suburbs) reduced aggregate
travel by a far larger 32%.[14]

Another evaluative benefit comes from comparison of the
highly disaggregated solution, with more aggregate ones.
That is, we may disaggregate the patients to permit Negroes
to visit only those hospitals known to accept them, or we may
not make the racial distinction, and permit these patients to
seek the 'nearest available' hospitals. The pattern of flows
may be dramatically shifted, and the travel distance reduced.
Here, although the model does not reveal the real complexity
of flows, the nature of the difference in pattern, and the magnitude of improvement when racial discrimination is overcome,
can be reasonably estimated. Similarly, we can contrast
patterns when the charity patient is permitted freer access to
the entire system, and if religious distinction of hospitals
did not exist. Such evaluation is vital, since for example,
accepting discrimination on the basis of ability to pay might
lead to a conclusion to build an additional charity hospital
whereas, if the discrimination were overcome, no such facility
might be needed after all.[15]

Gravity - Interactance Models

Another time-honored, yet much-abused hypothesis of spatial
behavior lurks underneath the 'gravity model' approach to flow
predictions.[16] We find the model frustrating, mainly because
it does a rather brilliant job of reproducing the structure of
flows, and yet we are not fully confident what goals people
have in fact pursued, and due to mathematical problems, do not
feel entirely secure in using results for evaluative purposes.
The interactance theory is not strictly optimizing. Distance
is not minimized, but the model is behaviorally rational.
Since greater distance diminishes the likelihood of contact,
the model tends to minimize distance. Prediction *may* be
optimal, *if* the model is considered to minimize the joint satisfaction of an heterogeneous population at minimum distance
but where satisfaction requires the use by some of the people,
or of individuals part of the time, of a variety of opportunities. A set of prospective patients viewing the opportunities around them cannot be expected to think alike: some
may consider a closest opportunity 'best' while others may
view a farther one better because it is larger. From an aggregate point of view, with people varyingly substituting size
for distance, various destinations may be equally good. The
decision-making hypothesis underlying the gravity model --
that the individual evaluates the probability of choosing
various destinations on the basis of size and distance, and
that a set of individuals will therefore tend to be distributed
by highest choice according to these probabilities -- is really
a rather good one. The motivation for each area to follow
its own interest is far more realistic than the 'system' optimum of the transportation problem. Also the individual and/or
group evaluation of opportunities can take into account differences in perceived attractiveness of destinations due to
religion via substitution between distance and religious

satisfaction, something far more difficult in the transportation problem.

Thus much theoretical criticism of the gravity model is unfounded. Still, the problems of actually mathematically constructing the model may be great. For example, we conducted several regression tests of the ability of the gravity-type formulations to account for the pattern of flows of patients between communities and hospitals. Most significant variables were community population, hospital size and number of physicians, other beds intervening between the community and the hospital; population intervening between the hospital and the community and degree of religious similarity of hospital and community, that is, a traditional intervening opportunities version of the gravity model. The results, with R^2's around .65, are really amazingly good, considering the fine level of disaggregation and the complete omission of known factors of importance -- the physician and means of payment.[17] Flows over paths, and paths used and not used, were fairly well predicted, and there was a logical behavioral basis for the model. However, as a complete model, the statistical version predicts the level of total flows as well as paths, from each area and to each hospital. We are not at all sure whether the divergence of these from the actual levels are useful as evaluative data, or pecularities of the mathematical construct. Ideally, we could interpret over-predicted outflows from an area as an indication of excess near capacity. Model results do confirm these expectations, but not always consistently, nor as elegantly as the transportation model can via the dual price structure. Overpredicted flows between given communities and hospitals show potential larger movement, evidently being restrained by discriminatory controls, and strongly underpredicted flows drew attention to perhaps somewhat irrational flows being forced upon some areas and hospitals by such restraints.

C. A Simulation Model of Physician and Hospital Use

The interactance model is seen to have an advantage in the area of how the individual evaluates surrounding opportunities, but is not too satisfactory in the actual allocation of patients. The 'transportation problem' has the advantage of a built-in evaluation of efficiency, but in this social context at least, is far too simplistic. The simulation model to be set forth attempts to utilize the advantages of both models.[18] It is an interactance model to the extent that probabilities of patients from an area visiting various hospitals are estimated from such a construct, though modified. But this stage only provides an initial estimate of hospital use. The model has two additional useful features: a mechanism to achieve replication of the system and one to reallocate hospital capacity in order to improve upon the present system.

Data on communities, physician clusters and hospitals are presumed available. Numbers of patients of a variety of characteristics and demands are known or can be estimated. It was observed above that there is little or no substitution possible between obstetric, pediatric and medical-surgical units of hospitals: similarly, the higher level hospitals have a 'monopoly' in caring for many kinds of cases. Thus if data permits division of patients into types and levels of demand, the simulation model would need to be run separately for significant combinations of type and level of care. At least six are needed: pediatric, obstetric and medical-surgical or 'high level' and 'low level' care.

The first stage (A) allocates *patients to physicians*. For each community, white patients seem to view the attractiveness of physician clusters as a simple function of the number

SIMPLIFIED FLOW CHART: PHYSICIAN AND HOSPITAL USE SIMULATION
(for each combination of level and type of care)

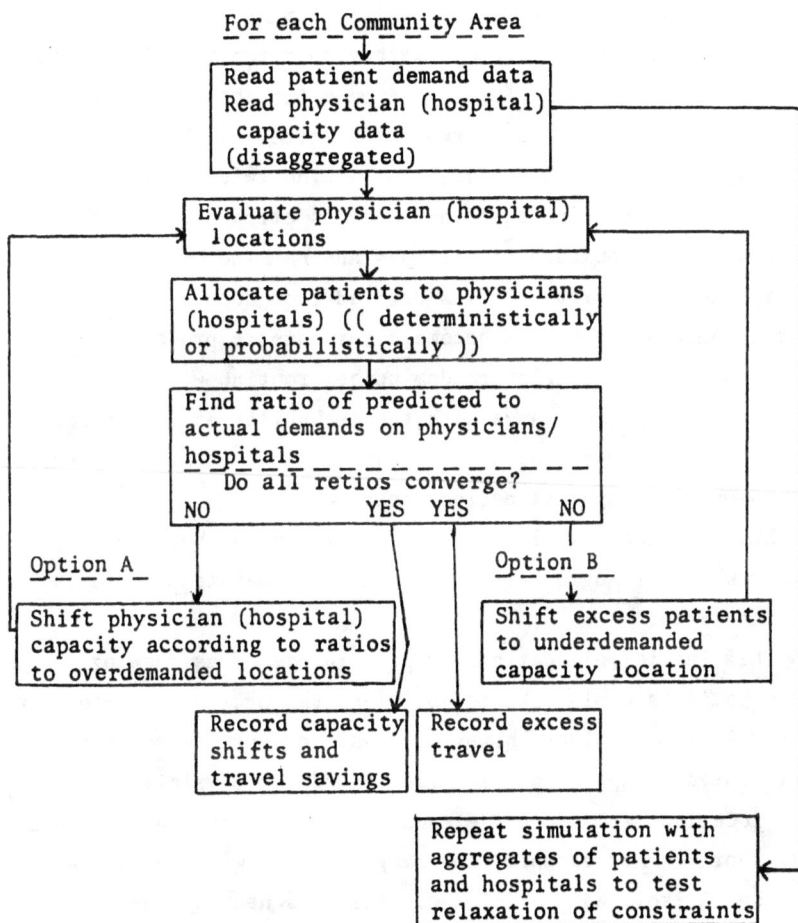

For each Community Area

Read patient demand data
Read physician (hospital)
 capacity data
(disaggregated)

Evaluate physician (hospital)
locations

Allocate patients to physicians
(hospitals) ((deterministically
or probabilistically))

Find ratio of predicted to
actual demands on physicians/
hospitals
      Do all retios converge?
NO   YES YES   NO

Option A   Option B

Shift physician (hospital)   Shift excess patients
capacity according to ratios   to underdemanded
to overdemanded locations   capacity location

Record capacity   Record excess
shifts and   travel
travel savings

Repeat simulation with
aggregates of patients
and hospitals to test
relaxation of constraints

Note: Since the procedure for allocation
        to hospitals is the same as for
        allocation to physicians, charting
        is not repeated. In fact, the
        hospital allocation builds on the
        results (Option B) of the physician
        allocation.

266

and variety of physicians (white only).   This is supported by
studies of trips to services in general.   But the likelihood
of patients visiting the clusters is influenced by the cost of
reaching them, the fact of closer intervening physicians and
the loss of information about farther opportunities.   Patient
reaction to distance, as derived from actual behavior, is one
of indifference up to about two miles, after which attractiveness falls rapidly -- that is, a physician twice as far would
be viewed as something less than half as attractive.   If
there are large numbers of patients and not too many potential
destinations the probabilities derived from such a modified
interactance approach can become deterministic proportions.
Otherwise a "Monte Carlo" random number routine will allocate
the patients in accordance with the probabilities.   A probabilistic approach may also be appropriate owing to the observed
'randomness' in typical patient behavior.

     Negro patients will be similarly allocated, but will visit
either white or Negro physicians, with a moderate preference
for the latter.

     This initial allocation may be interpreted as showing
where patients would like to go, given the present distribution
of physicians.   Since, however, physicians are not evenly
distributed among the people, the demand for physician care at
some locations will surely exceed their capacity and at others
fall short.   For example, too many patients will be allocated
to a small isolated physician cluster in a heavily populated
area.   According to the mathematical operation of the modified
interactance prediction of flows, a large cluster of physicians,
with competing clusters nearby, will not attract enough.
Since a consistent basis for choice is applied, we can interpret the difference as a measure of the imbalance or inefficiency of the location of capacity.

Substage A_1 *shifts physician capacity* until the demand on each physician cluster comes within some acceptable range of divergence. For example, assume the initial demand for one physician cluster is twice their normal capacity, and on another, half. Presumably, there are too few physicians at the former, too many at the latter. As a guess the model doubles the number of physicians at the first, halves the number at the second. The allocation, A, is repeated, with the altered capacities, and the divergence between normal and predicted demand rechecked. After four interactions of this procedure, the demand comes within an acceptable level of divergence (in our example operation, ± 10%.)

Substage A_2 conversely *shifts patients* from overdemanded physician clusters to underdemanded ones, in order to replicate, as closely as possible, the actual pattern of travel. After the initial allocation, flows to overdemanded clusters are proportionally reduced to actual capacity. The residual demand is then reallocated to underdemanded clusters only, a procedure which also requires reiterative allocation. The greater aggregate distance that is required will be a direct measure of the inefficiency of capacity location: that is, an estimate of the extra effort patients in fact must exert to get care. Comparison with the shifted-capacity solution measures the savings possible from such relocation.

The second state (B) allocates *patients to hospitals*. The replication of actual patient-to-physician flows are taken as inputs. In accordance with earlier discussion, it is necessary to divide the patient population into six subgroups: paying Negro patients (as above); charity patients -- those who did not visit a physician at all, but will visit hospitals directly; and four white paying subgroups: Jews, Protestants, Catholics and the religiously indifferent. Since we know the capacity of hospitals to care for Negro and charity patients,

each of these allocations is done separately. All white paying
patients are allocated in the same model run, but the religion
of the patients is recorded. Within the white paying group,
allocation to hospitals is seen to reflect a double substitution: a balance between distance to hospital, size of hospital, and its religious character. Within the Negro group
and charity group the balance is just between distance and
size.

Charity patients are allocated to hospitals in the same
manner that patients were allocated to physicians above -- as
a simple function of distance to hospitals and their capacity
to treat charity patients. For Negro patients, allocation is
as before, except that the probability of visiting a particular
hospital is a function of both the patient's and the physician's
evaluation of size and distance. The working hypothesis here,
that the choice is equally a function of distance from patient
and from physician, seems more reasonable than that one or the
other's desires should be controlling (in effect, the mean
distance to the hospital from physician and patient is substituted for distance from the patient).

White subgroups evaluate distance "religiously" as well as
geographically -- that is, a mental barrier is placed against
a hospital operated under the auspices of a different religion,
which increases the effective distance to it. Analyses of
actual flows and experimental operation of the model suggest
that on the average Jews evaluate the distance to non-Jewish
hospitals as about three times farther; Catholics evaluate
distance to non-Catholic hospitals as about twice as far;
Protestants evaluate Catholic and Jewish hospitals as about
twice as far, but evaluate nonreligious oriented hospitals
about the same as Protestant hospitals. These factors,
applied to the distances to hospitals, effect the probabilities
of visiting various hospitals; otherwise the allocations method
is the same as before.

Substage B_1 then *shifts hospital capacity* until the demand
on each hospital comes within some acceptable range of divergence (for Negroes, charity patients, and white paying patients
separately). The same iterative procedure as in Stage A_1
shifts beds from underdemanded to overdemanded hospitals.

This substage of the model has the capability, if desired,
of creating new hospitals and estimating their ideal size.
Plausible locations are given a 'token' hospital of but one
bed; if these locations are superior, beds will be shifted
from poorer existing locations. Some present hospitals may
in fact be eliminated by the model, although it is also possible to prevent an uneconomic reduction in size of existing
hospitals.

Substage B_2, as A_2, shifts patients, again for paying
Negro, charity, and white paying patients separately, from
initially overdemanded hospitals to underdemanded ones.
Again, the greater aggregate distance traveled measures the
extra effort patients must exert, given the present distribution of capacity. Comparison with the shifted capacity solution measures the savings attributable to relocation.

Since all the disaggregated flows will have been allocated,
it is then possible to summarize all flows and demands on
hospitals, and to make summary comparisons. Addition of
initially predicted demands on hospitals on the part of all
three subgroups will yield net measures of capacity imbalance.
Summing of the suggested capacity shifts will indicate net
shifts both as between locations and as between subgroups --
for example from white paying to charity patients.

The third stage (C) of the model is *experimental*. The
substages of initial allocation, shifting capacity and shifting
patients are repeated for any desired 'external' changes.
For example, several new hospitals or expansions may be

approved or anticipated. The effects of such planned reloca-
tions on aggregate travel of patients and on the demand for
existing hospitals may be measured. Estimated populations as
of some future date may provide the basis for estimates of
patient demand by area. The resultant imbalances and suggested shifts become a valuable indication of where new hospitals or expansions are needed.

A particularly valuable experiment, given present demands
and capacities, measures the effects of *relaxation of constraints*. It may not be necessary to carry out all the shifts
suggested in the model in an attempt to meet the separate demands of the many subgroups. Certainly it would be so costly
that only some portion of suggested relocation or new capacity
would be justified. Thus it is important to discover whether
it would in the end be cheaper and easier to relax some of the
present restrictions on entry to hospitals. Some or much of
the apparent imbalance might disappear if patients and physicians had freer access to the system. This can be tested
by appropriate aggregation of subgroups. The most feasible
changes are in regard to race and ability to pay, since legal
and financial arrangements can be made to permit entry to
hospitals irrespective of color and income. Since preference
on the basis of religion is personal, relaxation here is somewhat academic, so long as hospitals under religious control
exist. Aggregations to be tested then are (a) all patients
irrespective of color (b) all patients irrespective of income
and (c) all patients irrespective of color or income. These
tests of relaxation of constraints may be applied to both the
patient-physician and patient-hospital stages of the model.

D. Results

Allocation to Physicians: Shifts of Capacity. As expected, the initial allocation resulted in excess demand on
physician clusters in newer and poorer areas, and insufficient

demand on older and larger clusters. Almost all demands were
brought within plus or minus ten per cent of the mean load
(eleven patients per physician) within five iterations. Some
1500 (about 15%) physicians were shifted, mainly from the Loop
(Central Business District) and other very large clusters to
smaller clusters closer to the population. Although this
shift greatly reduces aggregate patient travel, we realize
that this break-up of agglomerations may be uneconomic. In-
efficient reduction of clusters can be avoided in the model,
however, by disaggregation of patients and physicians by major
specialty group.

Shifts of patients: when patients are forced to use the
existing system, total travel reasonably approximates that
actually observed. Aggregate travel exceeds that for the
physician capacity shift solution by 90,000 patient-miles, or
about twenty per cent. These model results are particularly
useful in pinpointing which specific areas and groups presently
incur the greatest excess travel. Not surprisingly, these
are paying patients in poor communities, and patients in
rapidly growing newer communities.

Allocation to physicians: race and income barriers removed. If patients, irrespective of race and ability to pay,
were able to visit physicians, an even greater shift of
physicians is forecast, mainly from the Loop and most wealthy
areas specifically to Chicago's poverty areas. This physician
shift is a measure of the great latent demand for physicians in
low income areas, or put honestly, of the unmet need.

Allocation to Hospitals: Effects of religion. For the
hospital trip, patients modify the distance according to religious preference. The model worked rather well in this respect, requiring, for example, greater average travel for
Jewish patients, owing to the limited number of Jewish affiliated hospitals. The capacity-shift portion of the model

also differentiated by religion. For example, in heavily
catholic southwest Chicago, bed complements of Catholic hospitals were increased, of Protestant hospitals reduced, in
reflection of demand shifts in the postwar period.

Shifts of capacity: The initial allocation resulted in
excess demands on hospitals in Negro areas and in many suburban
areas, and insufficient demand for inner city hospitals and
charity and veteran's institutions. The model results suggest a shift of over 12,000 beds (about 16%). Beds for both
paying and charity Negro patients are shifted to ghetto area
hospitals, at the expense of close-in hospitals, especially
Cook County Hospital (charity). Beds are added to many suburban hospitals in rapidly growing areas. Many hospitals on
the Chicago north-side are reduced in size, reflecting long-
term population shifts.

As with trips to physicians, level of hospital care was
not explicitly treated in these first example runs. Thus
Chicago's best and largest hospitals are slashed in size,
since they are indeed too large and central with respect to
general levels of care. In later model runs, a separate solution will be obtained for cases which could be handled only by
larger hospitals enjoying scale and agglomerative advantages.

Shifts of patients: If patients are again forced to use
the existing system, an excess of 116,000 patient-miles,
(about 20%), is required, over that when capacity is shifted.
The majority of the excess travel is incurred by black and
poor patients generally, since they are presently restricted
to so few hospitals.

Allocation to hospitals: Race and income barriers removed. When the barriers against free use of hospitals by
paying Negro patients are removed, patients' demands are
ideally met through the shifting of but 1380 rather than 1700
beds to ghetto area hospitals. Likewise patients are forced

to travel less far, given the present distribution of capacity.
If the barriers against free entry to hospitals by charity
patients are also removed, patients using the existing system
enjoy great travel savings, and a far less radical and therefore less costly shift of beds would be required to achieve
the same improvement in patient travel and hospital utilization
(for example, Cook County Hospital is reduced from 2700 to
1590 rather than to 800, and altogether 8650 rather than
12167 beds are shifted, for an almost identical travel savings.)

E. Conclusions

The simulation model outlined above is intended both to be
able to reproduce an actual pattern of use satisfactorily, on
the basis of properly understanding and formulating decision-
making criteria, and by extension, to evaluate the imbalances
of capacity at present, and to estimate the shifts necessary
for desired improvement -- to patient, physician and hospital.
A fair degree of complexity was required in order to depict
the system realistically. A partly deterministic and partly
probabilistic simulation model resulted, since the range of
choice confronting the residents of an area seems too great
for deterministic assignment.

The model is working moderately well for replicating use
of the system, at evaluating its locational efficiency and
suggesting shifts in location and policies that would raise
the general level of satisfaction at the least dislocation.
On the other hand, we note these problems: (1) The model
results are quite sensitive to the particular parameters of
the equation (patient interpretation of distance, size and
religion). Hence we cannot claim that the results are
'right' until we obtain much more evidence, including personal
interviews, of patient perception and behavior; (2) we are
not fully satisfied with the specific mathematical operations

of the model; (3) the value of our present results is limited by lack of breakdown by physician specialty and level of hospital care; and (4) it may be possible that this model places too much stress on reducing patient travel and not enough on the viability and quality of institutions.

Although the model was developed for the hospital use context, the programming is flexible enough to permit a much broader application at least to problems involving person movements, differential location of demand and supply and the evaluation of the efficiency of travel patterns and supply patterns. We believe the evaluative portion to be the most important contribution, although we are not yet confident as to how good it will be. If it proves useful, then further application to movements to shopping, schools, churches, recreation sites and others would be appropriate and indeed necessary for any demonstration of true generality.

NOTES

1. This research is part of the "Chicago Regional Hospital Study," and is funded by National Institute of Health grant HM 00452-01, and is co-sponsored by the Hospital Planning Council and the Illinois Department of Public Health, with the participation of the Center for Urban Studies and the Center for Health Administration Studies, University of Chicago.

2. G.D. Rosenthal, "The Demand for General Hospital Facilities," *Monograph 14.* Chicago: American Hospital Association (1964).

3. Richard L. Morrill and Robert Earickson, "Hospital Variation and Patient Travel Distances," *Inquiry* (1968).

4. Richard L. Morrill and Robert Earickson, "Variation in the Character and Use of Chicago Hospitals," *Health Services Research* (1968).

5. Gunnar Olsson, "Central Place Systems Spatial Interaction, and Stochastic Processes," *Papers, Regional Science Association* 18 (1967), 18-45.

6. Richard L. Morrill and Philip Rees, "Influence of the Physician on Patient to Hospital Distance," *Chicago Regional Hospital Study* (1968).

7. See Charlotte Muller, "Income and the Receipt of Medical Care," *American Journal of Public Health,* 55 (1965), 510-521.

8. Robert E. Coughlin, *Hospital Complex Analysis: An Approach to Analysis for Planning a Metropolitan System of Service Facilities.* (unpublished Ph.D dissertation, University of Pennsylvania (1965)); Jerry B. Schneider, *Planning the Growth of a Metropolitan System of Public-service Facilities: The Short-term General Hospital.* (unpublished Ph.D dissertation, University of Pennsylvania (1966)).

9.   Pierre De Vise, "Slum Medicine: Chicago Style," *Working Paper IV. 8*, Chicago Regional Hospital Study, (May 1958). Chicago Commission on Human Relations, "Negro Physicians and Medical Students Affiliated with Chicago Hospitals and Medical Schools," Chicago (1966).

10.   Richard L. Morrill and Robert Earickson, "Influence of Race, Religion and Income on Patient to Hospital Distance," *Jewish Federation of Metropolitan Chicago* (1966).

11.   Benjamin H. Stevens, "A Review of the Literature on Linear Methods and Models for Spatial Analysis," *Journal of the American Institute of Planners*, 26 (1960), 253-259.

12.   W. John Carr and Paul Feldstein, "The Relationship of Cost to Hospital Size," *Inquiry*, 4 (1967), 45-65.

13.   Robert Earickson, "Simulation Model of Non-White Hospital Use in Chicago," Working Paper III. 3, *Chicago Regional Hospital Study* (1967).

14.   This general procedure was used: beginning with the worst-off patient areas, substitute the shortest path they could take for the long predicted one; shift physician capacity accordingly, until all very long paths were eliminated.

15.   Robert Earickson, "The Case for Decentralizing Cook County Hospital: Some Applications of Linear Optimization," Working Paper III. 4, *Chicago Regional Hospital Study* (1967).

16.   Gunnar Olsson, *Distance and Human Interaction: A Review and Bibliography* (Philadelphia: Regional Science Research Institute, 1965).

17.   Richard L. Morrill and Robert Earickson, *op. cit.* (1968).

18.   See also, Robert Earickson, *Spatial Interaction of Patients, Physicians and Hospitals -- A Behavioral Approach.* (Unpublished Ph.D. dissertation, University of Washington (1968)).

For Product Safety Concerns and Information please contact our EU
representative GPSR@taylorandfrancis.com
Taylor & Francis Verlag GmbH, Kaufingerstraße 24, 80331 München, Germany

www.ingramcontent.com/pod-product-compliance
Lightning Source LLC
Chambersburg PA
CBHW070609270326
41926CB00013B/2477